BEGINNING'S END

Shaykh Fadhlalla Haeri

KPI

London and New York
in association with Zahra Publications

First published in 1987 by KPI Limited
11 New Fetter Lane, London EC4P 4EE

Distributed by
Routledge & Kegan Paul
Associated Book Publishers (UK) Ltd
11 New Fetter Lane, London EC4P 4EE

Methuen Inc, Routledge & Kegan Paul
29 West 35th Street
New York, NY 10001, USA

Produced by Worts-Power Associates
Set in Baskerville
by HBM Typesetting Ltd, Chorley, Lancashire
and printed in Great Britain
by Redwood Burn Ltd

© ZAHRA PUBLICATIONS LTD 1987

No part of this book may be reproduced in any form without permission from the publisher, except for the quotation of brief passages in criticism.

ISBN 07103 0220 7 (HB)
ISBN 07103 0221 5 (PB)

Acknowledgements

This work could not have been possible without the sincere and loyal efforts of the following people:

Preparation of transcript   Hajj Mustafa Shawqi, Ali Hassan Conway, Jamila Xaigham, Halima White, Muneera and Dina Haeri

Editing of text   Caroline Whiting

Typescripts   Khadija Amatu'llah Herman and Lorraine M. Stemp

Overall development of text   Batul Haeri Mazandarani

Contents

Acknowledgements	v
Foreword	ix
Introduction	xi
I. The Path of Dynamic Submission	1
1. The Seeker	3
2. The Chase	24
3. The Veil	35
II. The Root of the Matter	47
Introduction	49
4. Emotions and Attitudes	53
5. The Concept of Justice	69
6. Time and Non-Time	77
7. Decree and Destiny	82
8. Body, Mind and Intellect	94
9. Noble Qualities	102
10. Rules For the Wayfarer	118

III.	*Beginning's End*	127
	Introduction	129
	11. The Wealth of Nations	131
	12. The Debt of Nations	161
	13. The Hope of Nations	172
IV.	*Postscript*	177

Foreword

We live in an age where man is an endangered species. Because the meaning of man is in danger, his environment, the earth, is also in danger. We live in a time that seems determined to make man the slave of the lowest aspects of himself. Because of high technology global danger is now different to what it once was. Our materialistic enslavements do not allow an environment conducive for men to grow. In this age, if man is to be restored in his splendour as a being of knowledge, he must set out on the path of dynamic submission in search of freedom through self-knowledge, as all men of the past and present have done. This is the path of the prophets, the path of the gnostics, the one and only path of the men who have mastered themselves and who recognized the freedom of no choice.

A brief global overview of the age we live in is presented in this volume to show that outward existence is only a mirror of man's inner state. That modern society has reached this condition is because man has allowed the lower aspects of his nature to predominate in life.

It is the prevailing system that is the illness of our modern age and that enslaves man to his lower nature. It is the method itself of this age that is its disease and its tyranny.

Here, we refer to another way, and by it man is not endangered; he is liberated, and that means life for those around him. It is a quick and direct way and it leads to knowledge of the self and, therefore, the freedom of man. It is another way, and it acknowledges that man can only find contentment and peace in this existence of opposites if he has inner knowledge of the one Reality that underlies all experience of duality. Material or physical reality is only a platform or starting base for the spiritual development – for otherwise by itself it is a dead end. The state the Western world is in now is beginning's end.

Introduction

All of us are students of self-knowledge in varying degrees of commitment and intensity. Some of us are concerned with knowledge of physical and biological matters; others with subtler, inner realms of meaning. But whatever form our search takes, we are all seeking the ultimate knowledge that has brought about our individual destinies. Our present moment, our present situation, constitutes our momentary destiny. This momentary destiny is the result of the interaction between the laws that govern existence and our individual wills, an interaction that results in a balance of these factors. The moment is the only reality of which we can be totally certain, subjectively and objectively.

The truth of 'now' is what human beings share, and 'now' is beyond time. It is not subject to the moment just past or the one to come. It is simply now – and the now that one experiences is the product of the past and the cause of the future. Now is both independent of and connected to past and future. This truth is the root of the dynamism of cause and effect and of the balance of duality in this existence.

Absolute, pure 'now' is that state of balance achieved when the pointer on the scale of duality is at the exact mid-point.

When this occurs, each side is in total harmony and balance with its opposite. Duality is nullified and only pure, blissful 'now' exists.

We are born into this world in order to grow in experience and wisdom and to recognize the one unifying principle of Reality. We may recognize it in a limited way – for example, when we see ecological harmony – or we may know it in a more pervasive way when we experience the overwhelming presence of the Creator of all.

However we look at it, we must still ask, 'Having recognized the unifying principle of Reality, what comes next in our search for the purpose of existence?' After we have gained a measure of wisdom, wealth, harmony and security in this life, we then ask, 'Is there meaning beyond this? Is the recycling of the body to the earth from which it came the end of all experience?' The majority of us avoid this question, or try to rationalize it away, instead of confronting it totally and fearlessly.

Everything acts in accordance with its own nature. For example, steel belongs to the earth as iron oxide. Because of its strength, we use steel to construct bridges. But the moment a bridge is raised, the steel starts moving toward its destiny, back to dust; thus, it starts rusting. If we want to preserve the bridge, we have to balance the natural inclination of steel to return to its source with our human desire to keep the bridge from rusting. We cannot, however, pretend that the bridge will remain for ever. Steel belongs to the earth, and it will fulfil its destiny; man belongs to his Creator, and wherever he finds himself, whatever he does, his natural disposition will lead him toward the knowledge of his Creator.

Experience is the meeting of two opposites. This meeting point is the human heart, for in it we experience love and hate, hope and fear, peace and agitation, wakefulness and sleep, security and insecurity. The heart of man contains the entire spectrum of experience; whenever we appeal to the heart, we find a common denominator in humanity. Otherwise, our interaction becomes transactional or hypocritical, based on economics, politics and other man-made disciplines. Although these disciplines have their place, they are limited because they do not lead us to the awareness that whatever appears negative in existence is really in harmony and balance with the total ecology. Because we do not know this, we see incongruity, imbalance, confusion and division in life.

The philosophy of our spiritual model is based on the existence of a merciful Reality that transcends time, encompasses Its creation from beyond time and in time, and creates out of love in order that It may be discovered. Man's purpose is to discover the one and only all-pervading Reality that encompasses his existence. When we are mindful of our purpose, we find a common denominator in everyone's aspirations, hopes, disappointments and problems. All of our experiences – positive and negative – are part of the process of self-knowledge. The root of our behaviour can be explained by viewing it through the binoculars of the divine unity of the one Creator.

This work is a collection of glimpses which reveal the unifying substructure that underlies diverse experiences and actions, which is, itself, the truth. It is presented in the hope of confirming to the sincere seeker that the ultimate discovery is that all creation, attributes and actions, stem from and are sustained by the one Source Whose nature is independent of creation. A taste of the vast ocean of the Oneness increases the thirst of the seeker. Ultimately, the boundary between the seeker and the sought, the questioner and the questioned, and the effect and the cause becomes so faint that one is immersed in the joyful intoxication of the complete beauty and absolute harmony where subject and object have merged.

I
The Path of Dynamic Submission

1
The Seeker

The term 'gnosis', ultimate self-knowledge, indicates a state that has been attained by certain individuals throughout the ages. The gnostic considers that all prophets and all truly awakened men and women have attained self-knowledge. The great master of gnosis, Ali ibn Abi Talib, was asked: 'What is gnosis?' He characterized it as involving purity, abandonment and the search for the one Entity behind multiplicity. This search leads to the awareness of man's inner poverty and basic nothingness, which leads to the further recognition of the completeness and perfection of whatever situation he may be in.

In every age, men of self-knowledge have existed who have known that the foundations of this knowledge could not be acquired intellectually. These men knew that self-knowledge is accessible only to those who are prepared to undergo a profound existential transformation in order to acquire mastery over themselves. The names differ – the *sanyasi* in India, the gnostic in the West, the monks of the Far East – but the path is one. In the spiritual traditions of the Near East, from the earliest dawn of civilization, we see the same quest for self-knowledge. It is known in Arabic as *tasawwuf*, or Sufism, and the follower of this Way is called a Sufi. The origin of this word is *safa*, which means purity.

As one embarks upon the path, one becomes aware that one's present mental and physical condition arises from a variety of factors, ranging from the genetic to the environmental. Some of these factors are inherited; some are acquired; most of them can be changed. Some of the subtler influences on us, such as changes in the radiation in the atmosphere, are barely detectable. However, our recognition of any outward factor, whether subtle or obvious, depends on its existence within us. For example, if we have no love within our hearts, we cannot recognize love outside ourselves. It is the same with anger, violence or insecurity. We are conscious of all these factors through a higher consciousness in each of us which unites us.

If this higher pure consciousness is taken to its conclusion, its foundation is the basis of the Unity of mankind. The state, therefore, that each of us is in, each in his or her own internal cosmology, is a result of the influences and factors we have mentioned. They intermingle and superimpose on each other to create what we call 'I'. From a scientific or causal point of view, our overall state is, therefore, perfect, since it results from a combination of actions and reactions and the superimposition of various layers of systems.

Separation and Unity

From the gnostic standpoint, there is no separation. The concept of separation exists only for the sake of illustration and outward experience. It unfolds a situation that is completely unified. My inner state is completely unified, but if I were to describe my overall condition, I might say that my back is aching from too much travel and my stomach is upset by the altitude. These conditions result from and are revelations of the natural laws of the universe.

We deserve everything that happens to us for we get what we deserve, not what we desire. What we deserve is decreed; not superstitiously by some supernatural power, but by the manifestation of Reality through a combination of factors, both obvious and subtle, resulting in the final state. Every system is governed by a set of laws. In the case of a falling stone, for example, the gravitational force is predominant. Other factors may also influence it minutely, but every system has a limitation that governs its bounds.

As we have said, our present state results from many factors. All of these factors superimposed, collected, are connected together resulting in one overall state. Yet, for the majority of us, our inward condition is at best confused. We react to our inner confusion by blaming our state on some external factor, such as the weather or the government. The more intellectual among us may write up long, complex dissertations about the cause of our current situation. The seeker's objective is to reach a state of awareness so that he or she sees the perfection of the state he or she is in.

The Desire to Know

According to the gnostic teachings, Reality, or Allah, wanted to be known; therefore He created. Thus the purpose of creation is to come to know. If we set out on the path of self-knowledge, then we can only experience growth and increase.

One of the names of God in Arabic is *Rabb*, meaning Lord. This word is related to the verb *rabba*, which means to bring up to its full potential. One of the responsibilities of the Lord, therefore, is to bring people to their full potential.

If our objective is to know, we have to begin with an understanding of ourselves and our immediate situation. If I examine clearly, for example, the reasons why I lost my job – a difficult boss, unstable market conditions, a move to new premises – the loss will be quite understandable. If we simply act as pure observers and remove from our hearts any subjective psychologizing, we are bound to see perfection in every situation – however personally detrimental or unpleasant it may be. If we consider any situation purely from the viewpoint of the creational laws that govern it, we will see the perfect harmony behind it. This does not mean that we should condone destructive behaviour, but rather that we should observe the absolute perfection inherent in any situation, agreeable or disagreeable. Once we have reached this state of awareness, we have made a start on the path of self-knowledge.

When we see greed, or any other emotion, arising within us – for example, in the course of a business transaction – our very witnessing of our state means it is less likely to afflict us. If we are spontaneously conscious of the anger rising in us, it is less likely that the anger will overwhelm us. Once we are able to recognize these emotions as they occur, we are less likely to be controlled by them.

Recognition of Bounds

The next stage in self-knowledge is the recognition of bounds. Every system exists within certain set boundaries. The simpler the system, the easier it is for us to observe the boundaries clearly. In the plant kingdom, for example, a severe frost may cause the end of a species or bring about some mutation. The human condition is more complex because we are able to stretch the bounds that apply to us. Our physical bounds are wider than those of other forms of creation, while our inner bounds are wider still. We may, for instance, allow anger and hatred to fester within us for years before they erupt. If we do not recognize the bounds of systems, we will create disturbance in an ecology that inherently contains its programs of checks and balances. The lack of restrictions in society today has resulted in a confusion that is a reflection of our reality. The prevalent laissez-faire attitude means that we no longer know where a thing begins and where it ends.

Nature itself extends courtesies towards everything within it. These courtesies are bounds and they will be maintained because the laws of creation are eternal. The cosmos began from a point of non-awareness in non-time. You and I also began from a point of non-awareness. We were not aware of our existence in the womb, but awareness grew within us.

There are certain laws of the universe, such as the law of gravity, that will never change. Another is the law of action and reaction. If we give love and fairness to others, we will eventually receive it in return. It may not happen immediately, for we are tested by ourselves in order to know the extent of the purity of our intentions. We are programmed to learn the truth. The more sensitive and connected we are, the more quickly the truth will unveil itself, and insight and cognizance will prevail.

Recognizing that the laws of creation are absolute is part of our growth. We have freedom of choice, for there is no such thing as an experientially fixed state. This would be tyrannical, and creation emanated from love. It emanated from One and is permeated by Oneness. However, the laws of creation do not change simply because we are well-meaning. The law of gravity did not prevent the arrows from striking the Prophet Muhammad in the battle of Uhud. The afflictions of the Prophet Jesus did not cease because he kept on singing the one and only song of Reality, irrespective of what happened to him, as had every other prophet or reflector of truth.

Each of us has the potential to reflect the entire creation. Our nature is to want to share; that is why we cluster in cities, nations and cultures. We basically gravitate towards those who reinforce our own experience and knowledge, and erect barriers against those who do not. Different cultures, nations and movements are all attempts to unite because most of us do not realize that everything is already united. We do not see unity because we experience everything in duality – we see life and think of death. We know poverty and wealth, hunger and satiety, love and hate. We all know these opposites, and we all want to see how everything interconnects. In other words, we are all seekers.

Everyone is a seeker of Truth but few know it. Everyone is in submission – the true meaning of *Islam* – for we are in submission to the ultimate experience of death. The only certainty we all have is that daily we are moving closer towards death. None of us wants to die and most of us are afraid of death. This, too, is a reflection of the Eternal Truth, for the Absolute, Higher Consciousness, Allah, is for ever. We have the seed of that truth in us; therefore we want foreverness. Our very desire for that foreverness arises from our perverted love of God.

It is not possible to move along the path of knowledge if we do not begin to recognize the bounds and treat them with courtesy and respect. People are now beginning to realize that the serious state the world is in today arises from our neglect of these eternally fixed bounds. The very fact that more than half the world today is in poverty whilst the minority suffer from overabundance reflects our abuse of boundaries. We claim to be scientific, yet we have arrogantly forgotten those scientific laws that govern the subtler elements of being human beings. If members of a family cannot live in harmony, how can they influence their neighbours to live in harmony? It is not possible. The sage knows that the vessel will ooze with what is in it. Show me the way a man lives and what he eats and I will tell you who he is.

Many of us attribute the desperate situation of the world to colonialism, capitalism, Zionism, communism or some other 'ism'. This situation results, however, from nothing other than failure to respect the bounds of reality.

The last few decades in the West have seen the erosion of the moral fibre that was behind the great experiment of establishing these countries. People no longer cultivate their inner awareness of these bounds and their meanings. Once we become conscious of our transgression, we see that every action has a reaction. Whenever we transgress, we eventually pay a price; there is no escape from this natural law.

Freedom and Bounds

In reality there is no freedom. If we could be aware of all that is happening around us, each action we take would be specific and appropriate to a particular moment in time. We do not have the capacity physically or morally to absorb the entire spectrum of Reality; however, the seeker of self-knowledge seeks to gain that capacity.

That is why the man of knowledge is in the right place at the right time doing the right thing. His knowledge comes from the recognition that there is one guiding principle that permeates everything and that everything is contained within its web. This web is tangible and we can know it if we want to. Its root is the unseen and its branches are the visible, physical manifestations.

So the reason why we are in such chaos today is very simple. We have allowed everything to happen in the name of freedom and 'open-mindedness' without placing restrictions on our behaviour. Everything has its bounds, but we think we are gods, not recognizing, as the Master Ali said, that we contain the meaning of Godness within us. God will not interfere haphazardly; He will not change the law of gravity, for example. Yet we in our arrogance may forget this and try to defy the immutable laws. We may be destroyed for we are subject to these laws and not above them.

The love of Reality for us is such that we are given the freedom to transgress the bounds so as to recognize the limits and respect them. Yet, it is because of these transgressions that we find our lives in such disarray. We end up being isolated, selfish, and more dependent on our bank accounts than on the relationships we have with our friends, neighbours and countrymen. The lives of our forefathers were grounded in relationship, and they lived far fuller and happier lives.

Look at us now – the average working man may struggle for the whole year so that he can take a two-week fishing holiday – yet he could spend all year fishing. It does not cost anything. We fluctuate from one madness to another in our confusion, from heated anger to emotional, romantic love, and we call ourselves a scientific society.

Modern technology has given birth to amazing new complexities, some of which we are renouncing as we discover that technology feeds on itself. If we do not begin to see what is happening in each one of us, we will continue to apply artificial values to situations, not recognizing that the entire world and whatever is happening in it is knowable, because it is within us. If the entire cosmos is not contained within us in a meaningful sense, how else could we conceive of it.

We can point to only one thing at a time. If we are angry, we cannot be calm at the same time. If we are anxious about ourselves, we cannot show compassion at that same moment. Because of our current state of disunity, we are far more superstitious. We blame mishaps on luck and constantly seek the supernatural, not recognizing that the supernatural is only the natural which is not understood.

We are constantly falling prey to one 'ism' or another, not realizing that what we want to know is contained within us. We have either been given improper guidance, or we are weak and improperly disciplined. Therefore, we do not progress in self-knowledge, which is the only knowledge that is going to give us balance and show us why there is this apparent conflict within us.

We experience one state after another. A state of expansion is always followed by a state of contraction. How can an economy continuously grow in one country in the world unless it shrinks somewhere else? If people are suffering elsewhere, we may not be fully cognizant of their plight. We cannot have the experience of life unless we have the experience of death. We cannot fully appreciate the solace of friendship unless we experience loneliness. The man who lives his life at one extreme can have his life balanced only by the opposite extreme. Thus, if we really want to know the meaning of freedom, we must know the meaning of complete constriction. This is what all spiritual paths, all true religions, have taught.

Ultimately, to reach the goal of self-realization, all men and women of knowledge have to go through a period of reflection and retreat. It does not have to be in a cave in the Himalayas or in some other remote location. It is possible for us to have, here and now, the same experience as the great masters of history.

However a quiet, remote environment is very helpful until the seeker is strong enough to retreat into his inner cave, even though he may be in the midst of the market-place. In reality, no place is more spiritual than another. It is we ourselves who create situations from which we then have to flee in order to reach some level of quietude.

Ignoring the Bounds

We say we are a tolerant society. But what this really means is that we have inadvertently allowed all kinds of limiting factors to emerge, both inwardly and outwardly. Nature will recycle those who have gone beyond its bounds. The disease AIDS illustrates how nature comes to its own rescue. Respect for boundaries is part of scientific law. But through our ignorance, we have transgressed this law and have rationalized our mistakes.

We all want to be free, but we do not realize that we have inflicted tyranny upon ourselves by not recognizing the wisdom of the duality of existence. From the moment you and I are conceived, we are subject to the opposites – health and illness, growth and shrinkage, life and death. This seeming confusion is reconcilable. If we want to know, we will come to know, and the more we want to know, the more we will know. But if we are insincere, our desire for knowledge becomes a game and can even lead us to such transgressions as pursuing the occult.

Another transgression common in today's society is an undue concern with the future. People generally become concerned with knowing the future because they are not content with the now. But if we peek into the future, we are likely to cause ourselves distress. If we recognize with amazement, however, the perfection of reality as we can experience it right now and as it emanates from a pure heart, we will not worry about the future, for time will cease to exist for us. Indeed, there are people who are completely and soberly drunk with the ecstasy of the moment.

We contain within us the Book of Knowledge, which was engraved in the womb of our mothers at the moment of conception. This all-encompassing inner book gives us ultimate knowledge – provided we want to look at it. If we do, we will find that the knowledge of bounds and the fear of transgression is within us. The most useable book we can read as a guide is the ultimate book, the Qur'an. The most perfect being whose footsteps we can attempt to follow is the last of the prophets, Muhammad. But we don't come to ultimate knowledge merely by reading a book or following a guide.

We came into the world alone and we will go out of it alone; in the meantime we are totally responsible for our actions, and with that ultimate responsibility comes ultimate freedom. But our society has failed to recognize even the normal moral bounds that have been common throughout the ages in every spiritual teaching. Instead, we have transgressed according to every system's standards. We have become so hypocritical that even our discussions about gnosis, or the inner journey, have become superficial. The reason for this superficiality is that we do not know where to begin.

We have ended up living in confusion and hypocrisy because over a period of time we have locked ourselves into a tight corner from which we do not know how to escape. We live under more and more layers of self protection and end up being completely isolated, even though we may be surrounded by people. The reason for the increasing popularity of skiing and gliding and similar sports is the illusion of freedom and escape which they bring. Everyone wants that freedom, so what is it that keeps our hearts from soaring?

Freedom of Inner Detachment

The word for heart in Arabic is *qalb*; the root of it is *qalaba* – to turn, to revolve unattached. This means it is not desirous or expectant, or attached to any direction; it turns freely. In our society we often equate detachment with irresponsibility. But correct behaviour is based on inner detachment accompanied by outer attachment and adherence to the laws of nature. Not keeping to these laws will cause us only affliction. Even if we do not pay for our mistakes when we commit them, we will eventually reap the fruits of our actions.

We must recognize our bond with Reality; we must see that we are totally connected and that there is no separation. You and I appear to be different because of our outer physical differences, but inwardly we are the same. It is this sameness that connects the entire race of mankind. If we do not understand this we will continue to look for a quick formula to resolve our difficulties because this is the age of instant solutions.

We are now suffering from its side-effects: instant wealth, instant happiness, instant friendship. How can friendship in the true sense grow instantaneously? That could only happen if it is based on the ultimate foundation which is the love and knowledge of Truth. Then we will find everyone in harmony because there will be true inner courtesy, not merely outer courtesy.

The truth is always there, yet we have inadvertently been seeking false truth, in the name of convenience or economic progress or whatever reason motivates us. If we do not act in accordance with the laws that govern existence, we will pay a great price. We will eventually suffer from our mistakes individually and collectively. It is inevitable, for every action has an equal and opposite reaction. The more we see this fact and the more we live in the present, the more we experience timelessness and the moment touches the infinite. At this point we are exposed to the knowledge of Reality and the system that constitutes it; so we recognize wisdom in Reality's diversity, and unity in its infinite Oneness.

When people experience this sense of timelessness, beyond expression, they are commonly said to be 'awakened' or 'realized'. All of us have the potential to see everything as it is but if we do not want this knowledge, we will end up gathering worthless information instead. If we are not schooled in the outer, how can we train ourselves in the inner? We cannot start the inward journey unless we begin with the outer. The inner moves outward and the outer ends up containing the inner.

We begin by wanting the truth and end up by being aware and totally responsible. This correctness is not subject to religiosity, but rather to the divine laws that all the prophets tried to express, culminating with the completed message of the Seal of the Prophets, Muhammad.

The Teachings of Muhammad

It is the teachings of the Prophet Muhammad that we are trying to follow in order that we may live better, happier lives now. The teachings of Muhammad, like the teachings of all men of knowledge of Divine Unity, are not separate from their lives.

Muhammad's life was based on spontaneous, dynamic submission. He was completely interconnected with and aware of his environment, of the past and the future. He lived in a non-time zone within time. He was subject to all the biological fluctuations, to wakefulness and sleep, and all the other experiences that every other human being is subjected to. He was the same as everyone else; yet he was different from everyone else. His similarity was based on the sharing of experiences and consciousness. His differences sprang from his indescribable degree of awareness, and his living of each moment in true submission, in a perpetual and spontaneous state of adoration and absolute glorification of the Creator. His reverence for all creation, his desire not to harm life or to change it adversely, led him to enhance whatever situation he found himself in. His life was an instrument of evolution, in a spiritual sense, not in a Darwinian sense; he helped everything along its path towards its unfolding, towards the realization of its higher potential.

Dynamic submission leads to this state; submission itself brings about unity, and dynamism makes one's life rich and blissful because that is the prescribed path of nature. Outwardly Muhammad was subject to all that we experience. He had moments of comfort and satisfaction, of apparent bewilderment and of reflection. However, a man who lived the moment totally had access to that non-time state which is utter peace; so his interior life was without doubt that of complete contentment, harmony and peace.

His outer life, however, reflected the law of opposites in this world – health and illness, acceptance and non-acceptance. Muhammad encouraged those who were following the true path and he admonished those who strayed from it.

Muhammad never allowed people to exalt him. He often reminded them that he was a mortal, born of a mother, like all other human beings. He lived in this world but he was of another world. He was intensely alive, dynamic, and scintillating, yet he said, 'If you want to see a dead man walking, look at me.' He was dead to ambition, to expectations, to attachments, to fears about the provision of material necessities, to personal anger and jealousies. But he was alive to the ever-living Reality and Its actions manifested in this existence. He was a true slave in perpetual, spontaneous adoration and worship, moving where the winds of his destiny took him, for he knew that all is from God, by God's grace, and all returns to God.

Muhammad's teachings grew out of his life, and what came through him was the message of the Creator, the Qur'an – the eternal message, applicable in all circumstances, at all times. Because his life and his teachings were one and the same, his example has been followed by millions throughout the ages.

There have been thousands of prophets throughout history from different cultures, different ages – prophets of plenty, prophets of scarcity, prophets who harnessed visible and invisible powers, prophets who ploughed the land and taught only two or three among the poor, prophets who travelled, and prophets who simply sat and spent much time in invocation. All of these prophets have been part of the divine plan. What is the meaning of prophethood and messengerhood, and why is the emergence of prophets necessary?

The Prophetic Teachings

A prophet is the culmination of a mutational event that manifests clearly as a link between the past and the future, the seen and the unseen, this world and other worlds, the Creator and creation. Prophets are beings who are genetically and environmentally able to transmit, in a humanly realizable manner, the purpose of creation and the path that leads to its knowledge and fulfilment. As teachers, they are masters of the self. Yet as masters of life their very lives manifest the perfection and the truth of dynamic submission. All the prophets were in submission to their Creator. They are transforming agents, the catalysts in the perfect formula.

Adam

All prophets discovered their submission to the one Reality with their awakening. The first mutational situation in creation was the rise of consciousness in man. The Prophet Adam was in total harmony with creation in a non-worldly sense. His state was one of total tranquillity and harmony, and primal, blissful existence. He began to question. The rise of the questioning of the ego, (desire and expectation) aroused man's consciousness. With that arousal came responsibility and the need for discrimination.

The next stage that Adam went through was the gaining of knowledge of what was good for him and what was not, which desires were useful and which were destructive. Desire for the knowledge of the spiritual path leads to a healthy drive in man, but desire for worldly possessions or material or physical relationships brings disappointment. If desire for worldly things does not bring disappointment in this existence, it will bring disappointment at some other time.

Then Adam learned the art of submission. But neither the art of submission nor man's awareness of it could have been developed without the awakening of man's higher consciousness. The faculty of reasoning was dormant until the light of consciousness shone. Prior to his arousal, Adam was in a state of non-awareness in the garden of pure bliss.

Adam learned that the all-encompassing Reality enabled him to commit errors, recognize his (Adam's) dislike of their outcome, and start afresh. Through his dynamic submission, Adam obtained the keys to heaven and hell. By accepting the truth of his nature, his Adamic nature, his human nature, which can lead him astray, as well as by accepting his divine heritage, he obtained the keys to conscious choice. The result of his choices could now be used to delineate the bounds of the path, to avoid transgressing these bounds, and to adhere to the path, thereby re-entering the garden with the key.

Noah
As a man of knowledge, the Prophet Noah saw that the actions of his people were so ecologically unsound that nature itself was going to revenge and redress the imbalance. Noah foresaw the disaster that was coming, so he started in the most rational way he could to avert it. He attempted to change their actions so that their correct behaviour would create a new wave to neutralize the impending disaster. But after 550 years of weeping for his people – the name Noah is from the Arabic word, *naha*, to weep, to cry or mourn – of imploring them to mend their ways, all that Noah could do was to save himself, his close ones and the animals of the land. Final destiny, once it sets in, cannot be changed.

Abraham
The great and glorious Prophet Abraham met his final test when there lingered the possibility of inner attachment to his son. Prophet Abraham was given this most precious object of his affection in his old age. But a subtle doubt arose within him about the totality of his submission to his Lord. The test that arose from that doubt was the vision to sacrifice his son, Ishmael. Following the voice of truth within, Abraham began the sacrificial act. Doubt rose again, this time disguised as the

voice of reason. Why should Ishmael be sacrificed? What fault had the son committed?

Abraham's test was not about sacrifice or his love for his child; rather it was about transmutation and transcending form. Would he be able to detach himself from his precious and innocent son for what he knew was the all-encompassing Reality? When the moment for the sacrifice came, Abraham's decision to transcend attachment interlinked matter and energy, which were already interlinked in the subtle realm, and manifested in the appearance of the ram.

Once the decision was made and he actually embarked upon the act, the physical sacrifice was no longer relevant. Once intention and action are united, the matter is as though done. However, there is a danger that man's weakness and subtle hypocrisy may set in and stop at the level of intention. It is not enough to love only with intention; intention must be manifested in action. The inner and outer must connect.

The Prophet Abraham was totally determined to make the sacrifice and it was only when the barrier of that determination was crossed that the actual act was transferred to the ram, and not before. If we wish to emulate Abraham or any of the prophets, we must fuse our pure intentions with our actions, our worthy desires with achievements. Otherwise we will live unfulfilled and disconnected.

Moses

The Prophet Moses went through a similar self-discovery. He was inspired to deliver the message of Divine Unity to the ruling tyrant of the land. The human side of him questioned and doubted his ability. He said, 'Send me with my brother; he can speak well,' for apparently Moses suffered from stuttering.

Moses did not know what to do in the face of the mighty powers of the magicians. But his Lord said, 'You have their signs and powers and they are from me. It is not you. Draw your hand out and it will dazzle them.' Bewildered and uncertain, Moses approached the enemy. Suddenly that total submission in unity took place and Moses' beingness entered into that interspace between the seen and the unseen, between the human tyranny before him and the absolute slavery to his Lord to whom he was so magnificently bound.

So the man of true submission is no longer in a state that we, lesser mortals can describe and explain. The man of unity has already placed his investments in God's vault. He has put his ultimate treasure (his life) into God's hands; he has utterly, unquestionably, totally submitted; he is a dynamic flux in this existence – from moment to moment, from day to day, up and down, well treated or mistreated, respected or disrespected.

With signs and powers bestowed upon him the Prophet Moses walked with the tribe of Israel. No sooner did they cross the Red Sea, however, than the tribe of Israel started craving manufactured idols. As soon as Moses left them for his appointed forty days with Reality on the mountain, they disobeyed him and plunged into habitual disobedience.

Moses immediately recognized that this too was the plot of the Creator. Nature's job is to purify, to constantly strip away in order for everyone to submit if they will. Moses recognized his anger was a product of his expectations – he was not free.

What could he do but be in dynamic submission? What could he do but expect the best from what he was asked to do? Yet, he was always aware that Reality constantly pulls the carpet out from under our feet so that we realize that we cannot count on any earthly stability. Such a man gives in to helplessness and then becomes free of it.

Jesus
The Prophet Jesus was the pure master of the light of abandonment. When he turned the other cheek, he was in a state of total negation; that is, there was no 'he' to be struck. Because he was in a state of utter abandonment, whatever struck him struck Reality. We are liars if we speak the words of the Prophet Jesus without being what he stood for. Because of the discrepancy between word and actions, Christianity has become merely a romantic notion and a utopian ideal. Jesus came to put back the spirit of the Judaic law into the dead letter of the law

which had become the order of the day. Some time after him, however, what was left was a spirit with no clear code or law.

Muhammad

Muhammad acknowledged all of the prophets before him for they all were from the same source; they all had the same light, the same vibrancy, the same transmission. But Muhammad was the last one and the book that came through him, the Qur'an, the book of Allah, had no discrepancies in it. The way of Muhammad is known to us, especially if we look at the successor he appointed and at those who have followed him for centuries.

When we dwell upon the life of Muhammad, we contemplate a brilliant star, the last star in the heavenly revealed knowledges, which heralded the beginning and the end. As far as Allah is concerned, there is no beginning or end because He is the beginning and the end. Muhammad's message encompasses all the other messages that went before it, both in its inner meaning and its outer code of conduct.

Prior to Muhammad's time, humanity had not evolved far enough, both materially and spiritually, to be prepared for all the restrictions, as well as the freedom, that the Divine Creator intended for man.

Three thousand years ago, for example, it would have been impossible to forbid the intake of fermented fruit juices. At that time people did not know enough about preserving, and in the hot desert climate, many food provisions fermented and turned into alcohol so quickly that it was difficult to avoid consuming it even though it produced harmful side-effects. *At a time when their diets consisted solely of easily fermented foods, they could not have been given the luxury of deprivation of consuming fermented drink.*

It took over 23 years during the lifetime of the Prophet for the message to evolve to its completion. It was then that the Prophet said, 'Now the *deen* (religion) is perfected for all mankind.' He did not ask the people to stop their lifelong habits overnight, nor did he expect them to do so.

However, we cannot behave as the people of Muhammad's

time did. We have no excuse for not doing what is right and avoiding what is wrong in its totality, for not allowing what is good for us and forbidding what is not. We follow Muhammad not because he overruled the other prophets and their messages, but because he updated and completed the way of life that is best for humanity at all times and in all places.

The prophet is a being who is infallible; he is in total peace and harmony, witnessing discord and dispersion while being fully anchored in harmonious contentment, as if in the eye of the storm. The prophets, those great men of perfection, occasionally behaved in a manner that displayed human frailty or emotionalism. This behaviour did not detract from their noble, divine state.

That frailty is, in fact, a manifestation of Divine Unity and mercy so that the rest of humanity can relate to them. Acts of human frailty are the doors of the sanctified castle through which we can look. They are not moments of weakness. They are the proof of the love of the Creator for all creation through His perfect interlinks.

The prophet is our guide; he holds our hand and shares our experience. He is subjected to all of the tyrannies of life for our sake. He is devoid of egotism and selfishness. He is egoless and selfless; therefore he experiences all.

The closer a prophet is to Allah, the more he may appear in the eyes of his followers to be in affliction. The Prophet Job, the prophet of patience, saw nothing other than the perfection of the Creator in his trials and afflictions. He saw nothing other than the exquisite, magnificent interlink between cause and effect. He brought himself into the interspace between energy and matter; therefore, all of his supplications were accepted. Whatever he asked for was done, for his supplication was the Creator's command. What he asked for had already occurred in destiny. Job, the man living in time, simply reflected that destiny by his supplication.

When we enter the prophetic path, we are basically acknowledging the possibility of our attaining that state of dynamic submission that results in death-life and life-death, in vibrancy in all circumstances. As a result of that indescribable bliss that does not belong to this world, we are total lovers and followers of the Prophet Muhammad. We are also followers and lovers of all the prophets, at all times, wherever they may have lived, in whatever culture, by whatever creed.

There is no god, but God

There is only one creed and that is, there is no Reality but Allah. Once we acknowledge that, then we will know how Allah manifests this teaching through the prophets. Then we will understand the meaning of, 'Muhammad is His Prophet'.

To stop short at 'there is no god but God', is to be in the most sensitive and vulnerable state, for ignorance has begun to be removed and one side of the coin of Unity has been revealed. But the coin cannot be taken without the other side being revealed, Muhammad is the Prophet of Allah. We will remain in loss if we have not stumbled across this treasure.

Through an innate knowledge or scholarly reasoning, there have been individuals throughout the ages who have reached the conclusion that there is an all-encompassing, unifying force that holds together the seen and the unseen, life and death, and the two arms of the balancing opposites. But hearing a song is not the same as singing it; reading the menu will not satisfy one's hunger. Acknowledging justice is not the same as being just; having a good intention is not the same as completing the action.

There is only one path and it is based on the One; it is the beginning and the end and it is beyond time. We can only experience this timelessness if we stop the noise in our heads and sit with full consciousness, completely alive, yet completely dead to clutter. Then we will be in a state where we can begin to see the reason why we are in trouble and how to move away from it.

The Seeker

Once we have recognized that anything that happens is perfection, we can begin to act rather than react. Most of our energy is spent in reacting to changes in situations and states, because we have invested interest in maintaining a conducive or desirable state. The reason there is so little acting is because we spend so much energy in counteracting change or in adjusting to new situations. This situation will continue until there is a neutrality in our actions and we begin to act for Reality's sake rather than following our own whims. When this happens, the outcome of our actions will cease to affect us because they will be the outcome of pure and free intentions.

There is no possibility of our constantly living happily and correctly unless we begin to see where we are transgressing. The reason for the rampant spread of crime in our society today is that there are no bounds. Western societies are characterized by outer abundance and inner poverty. The societies of the past were often those of inner abundance and outer poverty. People lived with the continual possibility of death in front of them.

Today we shun any remembrance of death. We do anything we like because we do not constantly see ourselves six feet under ground. The way of unity is to see oneself in the tomb with each flicker of the eye.

If we choose, we can embark on the path of dynamic submission toward self-knowledge. It is for us to act. The laws of reality will not change, but we have the freedom to recognize them and unite our intentions with those laws. If we do, we will progress. If not, we will stagnate and degenerate. We cannot refuse responsibility for our actions for this implies ignorance. We do not want to be ignorant; we want to know.

2

The Chase

We are all looking for a stable foundation on which to build our life experience. We want a reliable foundation for our homes, our communities and our countries. We particularly want a firm foundation for the family because that is the basic unit on which all societies are built.

Intentions and Actions

But our foundations will be solid only if they spring from the one true foundation that existed before we were created and will continue to exist after we die. Because we have different backgrounds and personalities, we each look for this one foundation in our own way. Once we go beyond our outward differences, however, we will find that we are all basically seeking the same thing, although our quest may take different forms.

Sometimes our outer differences are so noticeable that we do not see the similarity in our real goals. But if we go deeply into the reason behind an act of violence, for example, we will find that it ultimately is based on love, although it may have been perverted to such an extent that it manifests in a destructive way. Excessive love of money, for example, might motivate someone to commit a theft. If we look beyond the act itself, however, we will ultimately understand how it all happened.

We all want to understand the underlying motives behind our actions. Each of us already possesses the root of that understanding. Its growth is dependent on its cultivation, exposure and nourishment. These factors govern the depth of our knowledge. Our search always begins with the quest for outer knowledge. We may start, for example, by creating a more suitable environment for ourselves, by having a better room to sit in so we are not disturbed by temperature fluctuations. But the time comes when we are satiated with physical comforts; then we become concerned with feelings. We want to share with people of like mind.

The time may come when we go so deeply into the one true foundation that we discover our total ecological connection with other beings and with everything that happens around us. Such a connection does not imply that there are no boundaries between entities or that we always understand everything. Rather, this connection means that our inner consciousness can fathom what at first appears incomprehensible; it makes our hearts and intellects sensitive to the incongruous or difficult.

We are meant to understand the external factors that affect us as well as our reactions to these factors. It takes an observer and an observed to create any situation in life. What we call 'I', the product of particular experiences and a particular background, meets an outside occurrence. The resulting interaction brings about a particular situation.

Men and women are expansive by nature. We cannot be contained by our bodily frames; we share on a much wider level through our higher consciousness. If we are travelling in another part of the world and let go of the prejudices – and other limiting factors – that we have been brought up with, we may suddenly find that we have a strong connection with a native of another country, regardless of his status, position or local culture. We may discover that he has fears and suspicions, loves and hates, insecurities and doubts, as we have. The objects of his fears may be different from ours. He may be afraid of the local despots becoming more aggressive, whereas we may be afraid that our government is not sufficiently protective of the small businessman. What we have in common is the experience of fear. The shape it takes may differ.

In a civilized environment, when fear arises in the heart we are taught to curb and repress it, expressing our concern through a newspaper article or support of a political faction. A peasant's fear, however, may lead him to wield an axe and demonstrate violently in his village. The peasant and the sophisticate share the same foundation, but each one's structure is erected differently.

When we discuss the way to self-knowledge, we start from the premise that we all inherently have the same gifts and operate within the same basic parameters. We are all conscious of what is right and what is wrong. Each one of us knows we have certain differences that manifest outwardly. They may be based on habit, environment, or genetic inheritance. Some of these factors may be beyond our control. They are the capital with which we have been endowed; our limitations in some areas may give us advantages in others.

Practically speaking, there is little we can do to improve our situation except to be aware of the tendencies we have. If I am aware that a change in temperature tends to make me more agitated, that very awareness is bound to enable me to be more stable. Likewise, I need to be aware of the effects modern technology may have on me.

One reason why television results in so much frustrated energy is that it is not interactive. When we see violence on the screen most of us want to stop it, but we cannot do anything about it. If we continually expose ourselves to negativity on the screen, we become more and more accommodating to concepts like violence until we finally cease to respond to them.

Breaking of Patterns

If we look closely at our outer differences due to cultural background, we will find they are minimal. One person may like his coffee sweetened; another may not. Many of our differences are merely superficial and based on habit. A willingness to change our habits would obliterate many of our differences, but most of us resent making such changes.

Indeed, one of the main requirements for embarking on the Sufic path is the willingness to turn away completely from what we consider to be essential for our well-being. Then we will suddenly discover how inconsequential these attachments are. When we recognize our dependence on our daily habits and voluntarily let go of these habits, we will find release from our self-inflicted bondage.

We are all the slaves of habit, even in mundane matters like brushing our teeth or combing our hair. We consider our own way of doing things the best way possible and this belief that our way is the best is the cause of a lot of our differences. Once we are able to perceive the real (habitual) cause and our unfounded attachment to our habit, the problem is sublimated.

If we put all these superficial differences aside and regard ourselves as creatures of the moment, remembering the possibility of death at any instant, then a vibrancy and urgency immediately come into our situation that push aside all the outer inconsequential layers, and we find that all human beings are basically the same. Then we will accept our outer differences and make allowances for the idiosyncracies of others, recognizing that we, too, have quirks of personality.

Ultimately we will reach a point when we want to know what the next step on the Sufic path is. We will have realized that we have to accept each other as we are and that we have a solid common denominator that we call higher consciousness. This consciousness lets us know deep down what is right and what is wrong; it provides a universal standard of morality.

As we have been witnessing in the last few decades, however, laws around the world have become lower in moral content and more hypocritical to accommodate the power structure, for our societies are based on materialistic rather than moral or humanistic values. The general trend in the world today is one of increasing decadence.

Yet throughout history the positive has always been found alongside the negative. As society moves towards the dishonest and dishonourable, an awareness of the need to strengthen the basic foundations and to re-establish unchanging values will increase. The real virtues of a thousand years ago are also applicable now and will continue to be so in another thousand years. They will not change.

The correct basic relationship between a man and a woman, their rights and duties to each other, for example, have not changed throughout the ages. Such rights and duties whether based on Christian values, or any other true religious values, are all basically the same. If we do not uphold these values, we will end up in a disastrous state, using such crutches as drugs and alcohol to cope with degenerating family life.

We all want to remember positive experiences rather than negative ones, good times rather than bad, for we were all born loving positive elements and wanting to avoid negative ones. Alcohol appears attractive because it helps us forget the past and lessens worry about the future. It makes us a little more spontaneous. The fact that spontaneity can be induced by alcohol means that its root is already within us. It is our responsibility as gardeners of the inner world to cultivate those plants, to develop our natural gifts rather than be dependent on harmful substances.

Why is it that I cannot admit my mistakes and forget my yesterdays, rendering myself open to the future? What keeps me from affirming that whatever comes I will do my best, according to my limitations, and live vibrantly, willing to pay the price for any mistakes I may make? The willingness to face up to one's mistakes does away with guilt, for it is guilt that plagues us if we are not willing to pay the price, whatever it may be.

Once we are fully prepared to accept the consequences of all our actions, good or bad, that thing called guilt, which causes a great sapping of our energy, disappears. We all have a certain amount of energy at our disposal, but that energy is easily dissipated by fear, guilt or unfulfilled expectations.

We have not explored the roots of our original foundation, so we have not dug back into the roots of our inner hearts to find out why we are doing what we are doing; our actions are only as good as the intentions behind them. If our intentions are pure and selfless rather than selfish and pleasure motivated, the outcome will be fulfilling. Eventually the rewards of our actions will be liberating.

The Unchanging Laws of Creation

The laws of reality are inscribed on the stones of the temple of truth. Time cannot change them. We may try to hide, cloak or distort them, but they are unchangeable. We are all seeking to live joyfully and happily, both inwardly and outwardly, but we cannot achieve that unless we are aware of the strengths and weaknesses of all the structures that we are connected to. We must not abuse our bodies by overstraining ourselves mentally, physically or spiritually, for if we undernourish any part, we will pay the price.

The majority of people in the developed world today are suffering from outer excess and inner starvation. We are full of inner loneliness, unable to open our hearts to other people. So we unburden ourselves to psychiatrists for a fee. The psychiatrist is like a man who is constantly repairing a roof whose basic nature is leaky. No sooner has he patched up one hole than another opens up elsewhere. The psychiatrist is not getting to the root of the problem. We cannot be dependent on him or on any other agent; our sole dependency is on Reality, that Entity that caused us to be. Whenever we develop a dependency on another human being, we generally add resentment, for what we really want are freedom and independence.

Although we are limited by our bodies and by our lifespan, our potential is unlimited. The more deeply we enter into any situation, the more likely we are to reach that strong bedrock upon which the truth is built. But, we have complicated our lives with so much outer sophistication that we find it difficult to dive into the ocean of Reality.

The remedy for our lack of inner nourishment can start by adopting a good or harmless cause, such as supporting an orphanage or feeding the poor. There is a great reward when we find poor people to whom we can give, for it is the giver who is the taker. But the more we learn about the spiritual path, the more we find we are reaching our fundamental bedrock. Then we have faith that our lives do not end with death. We become certain that the Reality that created us is infinite and that we return to it at the end of this life's journey. We know that we have been supported by its grace and nourished by its love and mercy. It is this belief that will help us to create strong families, neighbourhoods and nations. Without it, we feel cheated and angry.

Anger springs from an individual's lack of fulfilment. Because he is not getting enough inner nourishment, he becomes dependent on outer stimuli that will eventually drive him to despondency and despair. When we see antisocial or incorrect behaviour, we can be sure it is rooted in an individual's or a society's deviation from fundamentals.

Today we are obsessed with outer wealth, finding ourselves inwardly bereft. There is nothing wrong with outer wealth as such, but it needs to be balanced with inner wealth. Otherwise we are like a ship that has a huge hull filled with goods but only a tiny mast. Our ship will topple over under high winds because we do not have a spiritual mast. The wealthy man has less chance to remember death and the fact that he has no right to impose his personal power.

We have no rights in this life, only obligations. We all have a basic obligation toward creation, and whenever we are dissatisfied with life, it is because we have not recognized the fundamentals upon which existence is based. There are certain laws that govern existence, and if we adhere to them, our lives will be fulfilling.

However, if we transgress them, aiming for short-term profitability, we will eventually pay a price for our shortsightedness. Investing in the short-term is the curse of our times. Our forefathers, who invested in the long-term, also benefited in the short-term. They laid out parks and gardens, realizing that if they did not live to enjoy them, their descendants would. It was that love and concern that connected them with life. The Japanese economic strength is based on far-sightedness and sacrifice.

Societies will be destroyed unless they adhere to fundamental values. This does not mean we should live primitively and do away with all our comforts. The question, rather, is, 'Am I the master of my wealth and possessions or do they rule my life?' If the situation rules us, it is because we have acted in ignorance of the boundaries of our systems.

After all, nobody wants to be in turmoil; nobody wants to be constantly in debt, dodging his creditors, erecting still more barriers between himself and the Reality that created him. We all need to recognize the laws of nature and the Infinite without falling into superstitious meaningless rituals. The expressive aspect of a practice is just as important as the philosophy behind it, for one cannot have a living ideology without a code of living. The two go hand in hand.

So the man who believes in the generosity of the Creator must express that belief outwardly. If he genuinely believes in his nothingness and is willing at any moment to give up this gift called life – because he recognizes that after this experience the next gift, which we call death, is more infinite – there will be a transformation in his life that will be reflected in his daily existence. This transformation will bring about a coherence in his moral conduct that will provide boundaries for a garden which he will create in the heart – a garden of meaningfulness rather than a plastic imitation of a garden. It is this knowledge that makes the spiritually awakened person have no fear of death.

As a young man travelling in India, I had an intense distaste for any strong odour. A park where I often walked had some beautiful lilies that I loved to look at. But nearby there was a stinking pond which I always tried to avoid. One day a companion dragged me to the pond saying, 'What you love has its seeds here. From this will come the lilies of the future. It is only a matter of time.'

Sometimes the freshness of someone else's perception enables us to see the cause, which hitherto we have been blind to, of its effect. I was running after the effect – the flowers – not recognizing that its cause was something that seemed on the face of it undesirable.

It is the same with societies. The basis of a moral society may seem unattractive at first glance; it may seem antiquated (usually called backward) to hold on to a bedrock of values for which we may be called on to make sacrifices. However, if we want to live in a safe home, we have to build it well, not shoddily, like many of the houses of today. If we want something that is constant and real, we have to imitate the unchangeable way of nature.

Realizing that the laws that govern the outer and the inner are interlinked will enable us to adhere to them even though they may seem to go against us. 'Speak the truth even though it is against you,' says an Arab proverb. We will find everything eventually comes to us even though at the time a situation may appear adverse.

If we have tenacity, the rewards will come – but not without the companionship and support of people who share this path and enable us to live within the bounds. Once we recognize that the knowledge of these bounds is engraved on everyone's heart, our sailing will be balanced and our mast strong. We will continuously call upon the Infinite Reality, upon our own higher consciousness, to gauge whether our action is correct. Are we acting out of self-interest or out of desire for pure service? Each of us must be our own judge.

The Qur'an says, 'Nay, man will be evidence against himself, even though he were to put up his excuses.' The Qur'an also talks about the day when there are no more excuses and every cell will bear witness, for every cell contains the whole story.

We are all subject to certain absolute laws whether we like them or not. Once we recognize this, our limited existence has the potential of being a glorious experience. We must regard our life as diminishing capital, respecting every moment, counting every breath.

Instead, we have become so isolated and insular that we do not recognize the infinite capital we have, contained in a finite situation called time. Time, as we know, is relative; when we are content, the moment stretches so we lose track of its length. On the other hand, when we are waiting for a train, or for the bank manager to approve a loan application, time moves slowly. Contentment and tranquillity relate to least (or infinite) time, whereas disturbance makes us aware of time.

Ultimately, we are all responsible for our own lives, although we may try to blame adverse situations on a bad husband or wife, the neighbours, society or the government. The government is as good as the people who support it. The people in the developing countries constantly complain about being ruled by tyrants. But they have deviated in their lives from the tenets of true moral conduct. They have inherited a code of conduct but have not guarded it properly, so they deserve the bad rulers they have. Now they have to struggle, sacrifice and regain their lost heritage.

In Western societies, where much is inherited from the hard work of previous generations, most are now born into comfort. But whatever begins with ease ends with difficulty, for if we have not paid the price for the ease, we will not recognize the value of the effort that was needed: 'Easy come, easy go.'

Everything is just; the justice of God prevails everywhere. In the next life, we are promised, there will be no more 'we', only the pure and simple recognition of what is called *ruh* – spirit. Its state will be as good as it was the moment it left this life. If it has been liberated in this life, then when it goes into the post-death, non-time zone, that state of freedom is perpetuated for infinity.

But it is no use talking about the after-life when we all want to live in the state of the garden here and now. We can cultivate a lovely garden only if we know and apply its true foundation. We must not simply adhere to outer restrictions as though to survive an inspection, changing them after the inspector departs. The inspector is within us. Each of us contains an in-built, cybernetic system from which there is no escape.

3

The Veil

The Sufic tradition provides the seeker with a complete guide to the major steps in the journey toward self-knowledge. It also answers a number of questions that arise once we have embarked on the quest for Reality. What is it that prevents us from seeing the hand of Divine Unity in life and Oneness in diversity? What keeps us from spontaneously acting according to it? Why do we continue to afflict ourselves with our attachments? We need to examine some terms found in Sufism that will enlarge the dimensions of our understanding.

The Seven Patterns of the Self

The word *nafs* is a key term in the Sufi tradition. The literal translation from the arabic is 'self', and, in fact, *nafs* contains the entire spectrum of meanings included in the English word 'self'. The Arabic speaker knows which sense of *nafs* is meant by the context. The words, *nafs* and *nafas*, derive from the same root, which means 'breath'. *Tanafus*, the act of breathing, is based on two opposite happenings: inhaling air and exhaling it. There are a number of verses in the Qur'an that call for us to reflect on the nature of the *nafs*, for example: 'And the self and Him Who made it perfect. Then He inspired it to understand what is right and wrong with it.' In our reflection we ask, 'Is the self inspired to transgress or to be pious?'

Every person contains within him two opposing elements – the movement toward transgression and the movement towards piety. But awareness comes through recognizing boundaries and through the knowledge that transgression will cause us only affliction and danger.

In the Sufic system, the spectrum of self is composed of seven degrees, ranging from the highest to the lowest. Shades of grey exist between these stages, for the divisions between them are not completely discernible or quantifiable.

The Commanding Self

The lowest level of *nafs* is called the commanding self. The name implies that this self commands one to do whatever comes to mind. Neither emotional, nor rational, nor intellectual appeals get through to people in whom this self predominates. They are totally without guilt and nothing will stop them from acting out their whims. This *nafs* is impenetrable, despotic and solidified in its selfishness.

The Blaming Self

The second level is the blaming self. This self occasionally questions its wrong actions. This questioning indicates a crack in the solidity of the egoism of the self, allowing a beam of light to shine upon its reality and to occasionally reflect.

The Creative Self

The third level of *nafs* is the creative and tolerant *nafs*. When we are in an artistic or creative mood, we do not have many fears or anxieties and are open to inspiration. From the Sufic viewpoint, this pleasant self is in danger because its very openmindedness threatens the laws of correct behaviour to which creation is subject.

It is the open-minded self that says all right to everything and that anything goes. Like mercury on a table, this self jumps in every direction. It is the 'why not?' attitude. It is like a man of seventy who, having never skied in his life, suddenly decides he would like to try it. He will probably topple over and spend months in hospital recuperating from his injuries. Although the inspired self may find itself in trouble, it can also foster hope because of its flexibility. Most people who embark on a spiritual path start from this level of tolerance and liberalism because they are willing to see their own folly.

The Secure Self
The fourth level is the secure, certain self. The Qur'an reminds the self that is in that state of certainty to turn to its Sustainer, to return to the knowledge which it was given before its creation and to return to the state it was in before it could understand time, to return to its source, to its Lord.

Security and certainty begins with trust. Through trust we come to know. Because we want goodness and contentment in this life, it seems natural to accept the hypothesis that these states are attainable. Otherwise why should human beings have these desires? At a given moment, we may be unhappy and in trouble, but have faith that eventually we will come to know the cause behind our situation and learn how to extricate ourselves from it.

Setting out on the spiritual path, the seeker begins with the trust that what he is seeking must be right and attainable, although he has not fully reached it yet. As he daily progresses along the path, he finds he is in greater equilibrium as the level of his self-awareness rises. There is more connection between his inner intentions and outer actions. His trust helps to increase his contentment and security, and he is more steady and stable.

So we begin the path progressing from random inspiration into inspiration that is based on a discipline and on a trust that we will come to know. We embark on a path that we know is going to benefit us both immediately and in the days to come. This knowledge must be based on an inner reality and trust; how else can we talk about an end that neither you nor I can perceive or conceive of?

The Contented Self

The fifth level is the contented self. This contentment is based on the knowledge that whatever happens is the best outcome (for it is real), for reasons we can or cannot see. We are content with the ups and downs of life; content even when illness strikes. We may not fully comprehend the entirety of our situation; we may not realize the extent to which we have overworked ourselves; we may not understand that the germs that attack us only speed up the recycling process of wasted cells or tissue. They never attack an organ that is in a good state.

Contentment does not imply passive acceptance. It arises only when we feel that we have done our best. We are not content when we know that others or we ourselves could have been more aware or done more. Failing to do our best indicates inefficiency. We do not like inefficiency because nature and Allah's Way is the perfect way, the most efficient way, and we all strive for perfection in whatever we embark on, even if we occasionally find excuses to stop halfway and blame our mistakes on other people or circumstances. There is always an inner urge or drive.

The anchored self, the self that is sure, is content that it will come to know. It does not know now because it has been viewing everything through coloured, thick, and cloudy spectacles. It does not understand the total picture, but sees each situation in the microscopic or microcosmic. In reality, however, we are each a microcosm containing the meaning of the macrocosm.

So the contented self is the self that begins the spiritual journey and commits itself to the undertaking; it will not stop short until it comes to know the cause of its existence. How else can we be content? Otherwise we have only the certainty of eventual physical death and of being left, after all the experiences of the average human life, as food for the worms of the graveyard.

The contented self matures with knowledge. We have all been given a light of consciousness, which emerges after the mind has been tethered. *'Aql* in Arabic indicates the faculty of reasoning. It is usually translated in English as 'mind', but a better translation would be 'intellect' or 'reason'.

The Arabic headgear is called *'aqqal*. It is actually one cord twisted into a double circle and put on the head. As an item of dress, it is a functional device used to secure a piece of material that shades the head from the sun, but its other role is to tether the leg of a camel, so the beast sits down and behaves itself.

The origin of *'aqqal* is the word that means 'to be tethered'; if we are tethered it is by the faculty of reason. This faculty of reason is within us all if we stop the mind and allow ourselves to be quieted. It is for this reason that those of us who are spiritually inclined want to reflect. We want to stop the so-called mind and go wandering off. Access to Reality begins when the process of contentment, in a positive dynamic sense, leads to the contented self. I am content; hence I see more clearly. I see the despot within me. I see the blameworthy and the inspired within me, and I see the highest potential within me. I see freedom and timelessness within me.

The Pleasing Self
From this contentment emerges an immense inner stability and wealth that leads to the sixth level, the pleasing self. If we are content with every circumstance and situation that occurs in our lives, we will realize, spontaneously rather than analytically, the complexity of precision and perfection that causes each situation to occur. We may not like what we see; we may not expect it; but we will see the perfect truth in it.

We may, for example, have had certain expectations about our child's ability or performance. In the event that he has not behaved as expected, we are disappointed. Once we see that we overestimated the child's maturity, our understanding of our miscalculations will bring about knowledge and contentment. This state of contentment and understanding will not prevent us from acting positively to rectify a situation, or from assessing the possibilities for action from a balanced standpoint.

To reach the stage of the pleasing self means that everything in existence that interacts with us is content with us, for if we are content then the reaction or reflection is that everything else is content with us. There is no separation. We become secure in the knowledge that no matter what situation we are in, ultimately we will reap a return from all our actions toward others.

Our actions are investments that will pay off, one way or another. The person who is at this stage is, therefore, in complete equilibrium because he is aware of what is going on within himself and is connected to the world. He is also able to see clearly how he will, in time, reap the fruit of all his actions.

The Perfect Self
The seventh and ultimate level of the *nafs* is the perfect self. This is the state of perpetual spontaneous awareness.

One is essentially pure consciousness. If that pure light is directed at the lower end of the spectrum it will only encourage and propagate baser energies. If, however, it is directed towards the pure spontaneous state that we all aspire to and whose potential exists within the amazing complex physical mechanism, we will recognize that the limitations we face in this existence are there only to bring about knowledge of the unlimited. Then our direction is clear, it being to taste the limitless within, and living with these physical limitations becomes the most wonderful experience.

Sometimes one drives the body beyond its limit of endurance, this is caused by misdirecting our continuous drive to go beyond limitation. The limitless is to be experienced within. The body is the limited 'take off' platform, we have to learn this subtle differentiation and apply our energy appropriately.

Once we realize the full spectrum of the various unknown areas within our so-called self and we begin to see it spontane-

The Veil

ously, our afflictions are likely to lessen. If at the moment anger rises within us, we see that it is an expression of disappointment at being deflected from achieving a desire, we are then more likely to understand our miscalculation, and our anger will probably subside.

This does not mean that the spiritual seeker does not get upset. He may be very angry when he sees injustice, but there spontaneously arises in him a mechanism that brings about a practical outcome. Is there anything he can do about the situation? Can he stop the man from beating the child? If not, how can he ensure that it does not happen again?

If we are living in an environment that is degenerating because of its abandonment of virtuous values, ultimately a time will come when we are obligated by our teachings and by the precepts of our Perfect Masters to leave it, because that community or neighbourhood along with whoever belongs to it is doomed. Indeed, the blessed Prophet Muhammad said that a time might come when each of his followers would have no choice but to take a goat and seek refuge at the top of a mountain. He meant that a time would come when the situation in the world would be so decadent and hypocritical that a man of knowledge and truth would want to get away from the chaos and confusion he saw around him because he would not be able to do anything to change it.

There is a story about a seeker who, as he was approaching a town one day, saw a man running out of the gates of the city in great anguish. The seeker asked what was wrong. The man replied, 'There is nobody in this town who wants knowledge, so flee from such a place'.

One of the great masters of Sufism, was asked to define the Sufic Path. He said Sufism had been a reality without a name, but now it was a name without a reality. At the end of his life, Imam Junayd was found weeping. Asked the reason for his tears, he said: 'I have roamed all over Baghdad [then considered the great city of knowledge], and I have not found one heart that is ready to receive what I am transmitting.'

This overall situation never changes. It is the same today. As we get older, we all conclude that quality in the world is deteriorating. Throughout the ages older people have shared this belief. All the great masters have said that their own time was the worst of times. They cannot all have been the worst, but as our knowledge broadens with age and experience, we tend to see more conflict and disturbance.

From the standpoint of Reality, however, this is not the whole story. We know from the law of opposites that the more darkness there is, the greater is the potential for light. In maximum darkness, the tiniest spark shines brightly. Today, for instance, if we spend a few minutes of our day helping others, everybody praises us because there are few who sacrifice any time at all.

One day, A'isha, the young and outspoken wife of the Prophet Muhammad, said something completely out of place. The Prophet told her that it was the *shaytan* in her speaking. In Arabic the word *shaytan* comes from *shatana*, meaning to be cast off or far away from the path. There can be no 'on' unless there is 'off', no divine light unless there is evil. Our creation is based on duality in order to see that opposites emanate from the same source, so that the bounds are known. We cannot have good without evil, dark without light. When A'isha asked the Prophet, 'What about your *shaytan*?' He replied, 'My *shaytan* has given up. He is in submission, in Islam.' He meant that as negative tendencies arose in him, he recognized them and, with his instant recognition, banished them. This recognition comes from a state of constant awareness. These varying aspects of the *nafs* are within us all but they can be improved as we progress along the path of knowledge, if we have a clear direction, guidance and adhere to the limits.

These models or states of the self are only hooks for the mind to latch on to, so we can say, 'This is my lower self, my selfishness, arrogance or vanity.' All of us possess negative qualities. The only difference is that the man of spiritual insight will immediately see his arrogance, vanity or selfishness and seek refuge in the Creator. He will recognize the negative tendencies within him and their destructive potential. If he is a businessman, for example, he will recognize that arrogance is

The Veil

one of the principal causes of financial downfall.

There is an Arabic saying, 'The mistake of a man of reason (and wisdom) is a big one because when he makes a mistake it is as large as his reason.' We have all seen examples of a man who has lived correctly and responsibly throughout his life but suddenly at the end he makes a terrible mistake that causes his total ruin. Such an occurrence happens because he is not on a real spiritual path with clear bounds that show him how to behave in every circumstance.

The greatest master of Sufism was Imam Ali, the son-in-law and closest companion of the Prophet. All Sufi paths except one connect with him. He lived the life of a man of outward poverty, choosing always to dress in a patched robe.

When he was elected as a leader of the Muslims, he responded to the request reluctantly and continued to live frugally. One day he visited the home of a wealthy man who had prepared a lavish banquet in his honour. He asked his host, 'You have cared very much for this life. Have you invested in the same way for the next?' The man said, 'I have a brother who loves you and imitates you. We will bring him to you; he will please you.' The brother arrived in a dusty patched robe and Imam Ali said to him, 'What a miserable condition you are in. Why do you dress like this?' The man answered, 'I love you and I am imitating you.'

Imam Ali replied, 'But you are not me. I am afflicted with governorship and I want to live in such a way that the majority of the people will have access to me. Most of the people's living standard here is like this and I do not wish to be above them. Also, I want to show that it is not the garb that you wear that matters; it is who you are and what you represent. You who have been endowed with this wealth and well-being should show your gratitude to the Creator and the environment that has enabled you to have it by dressing as well as you can.'

The great-grandson of Imam Ali, Imam Jafar as-Sadiq, who is one of the pillars of our teachings, was wearing fine clothing one day, when he was approached by someone who questioned him, 'Your great-grandfather Imam Ali, who was our greatest master, always wore a patched robe. Why do you wear such fine clothes?' The Imam replied, 'Am I not dressed in garments quite commonly found in the marketplace of Madinah?' The man agreed.

Imam Jafar then said, 'As a master of the people, I like to wear what is available to the people. I do not like to exalt myself by calling attention to my dress (by wearing a patched robe or any special dress). However, if it was left to my personal preference, I would be wearing what I am wearing underneath this fine robe.' And he lifted up his sleeve to expose a threadbare, yellowed robe.

The *nafs* patterns we have been discussing exist in us all, but the more we dwell unnecessarily on the limitations of the level we are at, the more we reinforce them. It is for this reason we find contemporary psychology to be of little use, for it only serves to highlight a problem that in reality does not exist. The *nafs* is like a thief; the more we see it the more it runs away. Where is our anger? Once we have seen it, it disappears. What about our irrational insecurity? The problems we had last year have disappeared; our current ones will also disappear in time.

These patterns of the *nafs* are the shields that veil our eyes from the eternal truth. You and I hide the one and only Reality, which dwells in us all. But the way to recognize the infinite truth is by the recognition of the limited self. This is the meaning of 'He who knows himself knows his Lord.'

So we start by recognizing what goes on in ourselves; seeing that all our higher aspirations can only be achieved by recognizing the lower ones as they arise. The further we go on the path, the more we see everything disappear except the perfect beauty and mercy of God that encompasses all. Then we will melt into the one and only network of Truth, and our spiritual life as opposed to a merely physical life will begin.

Each one of us must choose whether to utilize our God-given potential. Time is short and our tendency is to postpone a decision. But if we invest a little of our time in the spiritual life now, that investment will blossom. If we dedicate a small proportion of our time to God's purposes, our investment will be amply rewarded. As our dedication and sincerity to the cause of Reality increases, so too, will the rewards. For our life is our investment; we are its portfolio.

Eventually, if we continue along this path, we will find that all these self patterns come and go like bubbles. Eventually what is left is the Truth that possesses, encompasses and permeates us. This can only happen if there are boundaries, for there cannot be freedom without limitations. If we do not know the meaning of constriction, how can we know the meaning of freedom? The more we are tethered, the more we are free, until we come to a point where we have absolutely no choice because we are living the moment. Then we taste life itself. Then and only then, we are qualified to talk about true meditation.

Having recognized our *nafs* pattern, everything becomes easier and lighter. As a result of that light-heartedness we begin to see the light of truth, for we are nothing other than the light. The Qur'an describes God as being the light of the heavens and the earth. That light burns in every heart, provided that it is a heart and not a bank vault locked behind steel barriers.

Each one of us has to discover how to live our own life fully. We cannot stop at second-hand knowledge. The spiritual path is for the adventurer (to add to the venture not to be reckless), for the one who truly desires, who has the right sensibility, who recognizes that with every breath he is moving closer to the end of this life experience. He wants to know life's meanings and his beginning. Once he knows the beginning, he knows his end. With that knowledge there will come a transformation that is possible for every one of us if only we are willing to abandon ourselves to it.

II
The Root of The Matter

Introduction

All of life's experiences are a search for the origin and foundation of our existence; for the seed of the human plant and what its fruit will be. You and I are based on matter, encapsulated in matter, and we want to know the basis of our being. What are its dynamics, and mechanics? What makes man tick? What makes him happy or unhappy, secure or insecure, content or discontent?

We are preoccupied with the forms through which we experience being. We want to find a standpoint that enables us to reliably recognize and understand these changes. Something in us seeks a secure viewpoint into the arena of the self. If we can get to the root of the self, if we can get to that point where we are both observer and observed, witness and witnessed, then we will experience the unity of root and branches with the soil and its vast ecology.

We are disturbed by the fluctuations of this life, by uncertainties both within and without. We can reduce our outer uncertainties by controlling our environment; we have mechanistic devices to control heat, humidity, and so on. But inner uncertainties present us with greater difficulties. For example, we have no control over our moods. So we don't know why we may become depressed. We want to get to the root of our violent mood swings because we want to experience stability. We seek reliability in friendships and relationships – but how can that be achieved unless reliability originates from within us? So the root of a situation is that fixed position from which we can observe its dynamics. Root implies the bedrock, the source and the cause of the matter.

The root of the matter is actually the point from which we see and understand everything that is in dynamic motion. It is the point of view of Unity. If we look at a situation from the point of view of Unity, we will find non-changing change, and that is the meaning in the Qur'an of, 'You will never find a change in the way of Allah.'

The guiding laws for every situation do not change. The things that change are the permutations, combinations, and apparent outcome, because they are in time, and time is change. There cannot be a situation or a time without change and there cannot be change unless it is in time. Although we are in time biologically, there is something in us that nags, as though it were a distant primordial echo telling us, 'You are not in time,' and therefore we want to stop time.

It is for this reason that when we have a pleasurable or joyful moment we want to capture it, fix it or immortalize it. This simple fact has created the enormously successful photographic industry. We want the moment of joy or pleasure to last for ever.

The root of the matter is that we are creatures in time, yet immersed and controlled in a non-time zone. Our behaviour reflects both a desire and movement towards fixation of time and yet through our intellect, our knowledge and our experience, we know that all of experience is motion in time. That is the puzzle.

In order for us to see this dual aspect, we must simultaneously be in a fixed position in the royal box of the gladiator's arena, seeing the self roaming around out there, and at the same time be attacked and suffer. We sit observing, while at the same time we are out in the ring with blows raining on us.

Introduction

What are the various emotions, tendencies and feelings that arise in us? Let us take these feelings and emotions one by one and put them in the gladiator's arena. Then let us take that royal seat, the fixed throne of absolute Truth, and observe these various emotions.

If we look at these emotions from an observer's point of view, as they arise in us, we will get a perspective different from being fear itself or being love itself, although we are that also. Otherwise, how do we experience love? How can we experience love unless it arises in our breast and we are simultaneously aware of it? How can we experience hate, how can we experience anger, unless both the experience of it and the awareness of it are taking place at the same time?

We would not be experiencing anger if we were not conscious of it. That consciousness of it is our seat in the royal box. The fact that we know we are angry means we are sitting in that fixed seat as well as burning with anger down there in the gladiator's ring. We are both the actor and the audience. We imply the awareness and the total experience. We become anger.

In order to catch every frame of the movie, in order to view ourselves frame by frame, we have to be in total submission in any given situation. If we are in total submission, the dynamism of the situation will be so slow that we will see our anger in snapshot stillness, blindly wreaking havoc. What I am talking about is both a diagnosis and a remedy – where there is illness, there is the remedy. Both illness and remedy are sublime. It is neither the illness nor the remedy, but the self-knowledge that matters. It is the magnificent, divine gift to creation – the Unific beingness.

From the fixed throne of absolute truth, then, we can view a variety of emotional and attitudinal states. From this point of view of Unity, we can see everything fixed, yet in dynamic flux. We can observe the causal pattern of the universal laws that govern existence, within its eternal timelessness, and we can come to understand the meaning of the divine decree in human life.

From this vantage point, we can see the integrity of the

human body, mind and intellect at work as we move toward a higher consciousness that increases our nobler qualities of character. And finally we learn to walk upon the path of dynamic submission, guided by the rules for spiritual wayfarers.

4

Emotions and Attitudes

Some of the emotions or attitudes that exist between the two poles of the human spectrum bring states of mind that are less than desirable. Although these states of being feel unpleasant, they present us with opportunities for growth and transformation. Let us look at some of these colourations.

Anxiety

Anxiety arises whenever we are confronted with a situation that does not conform to our expectations. We experience pressure. We feel conflict because our expectations do not coincide with reality. Anxiety can also arise from fear – particularly fear of the unknown. We are fearful that we may not be able to cope with a specific situation, that we may flounder, or we may fear inability to fulfil our obligations.

As a result of this anxiety we often dissipate energy and undermine ourselves and even bring on catastrophes. The remedy for this situation is trust. One way to develop trust is to recall the times we have been in difficult situations and to see how we have survived them and how they have helped us to grow.

Anxiety is also caused by the recognition of our inadequacy. For example, if I know I am inadequate in certain respects I will realize that things may not turn out as I want them to. At a certain point my weakness will show. However, the anxiety is not caused directly by my recognition of my inadequacy, but by the fear of not fulfilling a certain objective. If I face my inadequacy, the problem could be tackled and solved by bringing the adequate or necessary means to solve it. It is the fear of the discovery of my inadequacy that causes anxiety.

Anxiety is a symptom caused by our state of being. We create it and, like a silkworm, we spin it off. Everyone's anxiety has its own specific form although the general characteristic is the same. Some of us are anxious about material things, some about being loved, some about peer acceptance. Our anxieties are as individual as fingerprints, yet basically they are similar. Anxiety can be induced by environment; if the people around us are anxious, we will be anxious too.

Positive, or healthy, anxiety moves us in a desirable direction. It can create awareness and awaken consciousness. For example, if we are anxious because the road is slippery, we become more careful. Our anxiety becomes advantageous. It moves us to consciousness of safety. Because we want to be safe, we do not transgress the boundaries. Such healthy anxiety enables us to improve a situation or to learn not to repeat a mistake.

On the other hand, negative anxiety can bring panic, chaos, confusion and uncertainty. The tree of doubt grows so fast that it can overwhelm us. A common negative anxiety can be called 'in case' anxiety. I do not want the job in case my boss is arrogant. I do not want to cook in case the gas runs out. I do not want to act in case I make a mistake or am punished for my action.

Emotions and Attitudes

To avoid such anxiety, we need to learn to discriminate between good and bad, between what we can do about a situation and what we can't do. We need to act on what we know is right and avoid anything that puts us in a position of doubt. The realm of knowledge is as vast as an enormous banquet. If we are anxious about what is going on at the other end of the banquet table, we will miss what is in front of us. Our anxiety will prevent us from seeing clearly and will cause our attention to become diffused. But if we focus on those things that our faculty of reason tells us are conducive and acceptable, then we will gain strength and make progress. The more we dwell in a no man's land of doubt or in an area in which we are unable to act or function, the more time we will waste.

Another anxiety we may experience is that of not meeting our own objectives in relation to certain situations. This happens when we find that the future situation we had painted in our imagination is not unfolding in reality. But if we find that we can change our objective according to the reality of the situation, then we will be content. Our anxiety will be removed, for it is we ourselves who weave our own web of anxieties, according to personal specifications.

The logical part of our mind can relieve anxiety. But that part of the mind that we may term the psychological mind operates on the basis of emotion; hence it has the potential to cause us harm unless it is properly channelled. To say that a house is badly built is a meaningful and constructive statement. The owner of the house may be grateful for this information. If someone says, 'I hate this house. The design is awful,' the owner will be either hurt or angry or both. Nothing useful will be achieved.

One negative anxiety that is particularly common throughout our culture is the fear of death. This fear is an expression of love of life. We treasure life both in ourselves and in others. It is ironic that it is often the doctors who are most afraid of sickness, but it is the positive side of that fear that makes a good doctor. In this instance, as in others, we discover that the root of a negative anxiety can bring about positive drive and benefits.

Beginning's End

Stress

Stress is an applied force or a system of force that tends to strain or deform something. Usually a point is reached when the object under stress snaps or gives way; the stress factor of a piece of metal is its breaking point. Whereas previously there was a bond, this bond is now broken.

Psychological stress is actually a form of anxiety. If one is anxious because of a sore foot, then one is informed that it will take two weeks to heal and the stress is reduced. The pain is still there, but the anxiety is gone.

Recognition of stress itself diffuses a certain amount of the fear and anxiety. By facing the stress situation, one discovers that its power is decreased and ultimately it disappears. Stress and strain in a relationship can be seen as a warning signal that tells us something is out of balance. It is our nature to move toward an equilibrium. If we are pushed to the edge of our tolerance, we are probably being warned to return to the safety of the middle.

Stress and strain force us to recognize our limitations. 'Look, I am now under great stress, so don't talk to me.' This statement indicates that I am attempting to remedy my situation. A plant that is used to a particular environment goes into a state of stress when it is moved to another location. It needs time to adjust to different soil conditions and to find out whether the new ground is sufficiently fertile and conducive to its survival.

Stress implies a lack of tranquillity, but it also implies a tension that can be positive. Stress can force us to focus and to channel positive energy in the right direction. The pressure stress causes can become transformative. Stress arises because of the uncertainty about the right match between expectation and performance. Recognition of stress reduces the fear by a certain amount while it pushes us towards something better. Facing the stress and understanding it are a step towards its resolution.

Suffering

Suffering arises from the inability to balance one's intentions with one's actions, one's inner with one's outer. We expect that our feet will move at a certain pace, but we have a weakness in our limbs. When the cold weather comes, our expectations do not match our actions. Hence, we suffer.

Suffering happens when the things that cause anxiety become realized and demonstrable. We can feel the pain and the manifestation of suffering. For example the loss of a beloved friend precipitates pain; we suffer from the severance of a relationship.

Suffering can result from either a definable or an undefinable source. Sometimes its cause is visible and obvious; sometimes it results from imperceptible, subtle fears and realities. It is often a mixture of anxieties, fears, insecurities, and disturbances, physical or mental. Suffering from a visible source is more easily understandable and remediable – suffering from hunger, from poverty, or other physical deprivation. Suffering that originates from a more internal source, like mental illness, is less easily shared by others. It is more personal and less collectively understandable.

Ignorance is a cause of great suffering. If I do not know how to adjust the heat in my room, I am likely to suffer all night from the temperature imbalance. If I do not know how to behave in public and I commit a serious *faux pas*, I will undoubtedly suffer. Ultimately all suffering is connected to ignorance. We all want knowledge so that we may avoid what we do not like and experience what we do like. Suffering can thus motivate us to emerge from the abyss of ignorance towards the highway of smooth and tranquil travel.

But knowledge too can bring its own form of suffering. If, for example, I am in a car and I know the tyres are worn out but the driver and the other passengers do not know, then I suffer from the anxiety that is brought about by that knowledge. There may not be an accident, but I still suffer because of my knowledge whereas the others are spared that anxiety.

There is also suffering from disconnection. If we know something but are unable to share it, to have others join in that knowledge, we feel lonely and isolated and severed. We may feel disconnected from other people, from a particular situation, or from something we had or wanted. We all want connectedness and dislike dispersion and diffusion.

The prophets did not suffer in the same way as we do. They were suffering for the cause of Reality because the people deviated from the path of knowledge. They suffered because of the ignorance of mankind. This is the meaning of the suffering of Jesus Christ. There is nothing more blessed for the man of knowledge than to be with others who wish to know or who know; otherwise he suffers from loneliness like a doctor alone in a house full of lunatics. If he does not see another doctor soon, he too will go mad.

The wise teacher tries to enlighten the ignorant and show them the way. But if he reaches a point when he realizes that knowledge is not being assimilated or wanted, he will stop teaching and leave. The teacher's suffering is a positive energy because he wants to help the ignorant get back to the path of abandonment and Unity. The ignorant one sees that the teacher's suffering is the result of his veering from the path, and it is a positive incentive to get back into the unified field.

Suffering is a sign indicating that we have deviated from our path. Physical suffering is a deviation from physical harmony, mental anguish from mental harmony. The suffering of all seekers is a positive factor because it is inspired by the desire to move toward perpetual dynamic unification. With every difficulty there comes the opportunity for resolution of that difficulty. Resolution begins with an understanding of how our difficulty arose and it ends when we discover how to relieve our suffering.

Sorrow

Sorrow is a combination of disappointment and sadness. I may hope to achieve something but find my hope thwarted; I am disappointed. I may desire a certain house, or car, or friend, or wife, or book, but someone else obtains it. I am sad; the gap or desire in me is not filled by the peg I thought might fit. So sorrow results from a lost opportunity. However, if we reflect upon our sorrow and learn from it, then it becomes a positive experience.

So sorrow implies lack of knowledge. Otherwise, why would we fail? Failure implies imcomplete information and thus a lack of harmony. But if we can turn our sorrow into a positive experience, it will not leave any psychological or mental bad taste. It becomes only existential and prescriptive.

Sorrow can become a form of suffering. Sorrow is real in that it results from lack of information; it is unreal in that we create it ourselves by negative thinking. If we miss an opportunity, then we must accept our loss, not dwell on it, and go on. Not knowing why we are alive is also a form of sorrowful suffering. Not taking action against evil is a form of suffering. We suffer for anything that is not right; we are programmed that way by our Creator. If we become attached to luxury living, our attachment causes suffering whenever there is slight discomfort.

Every attachment is potential suffering; that's why our deepest inclination is to break it. Unless our attachment is for the love of truth, unless it is part of the provision on the journey, we suffer. Thus we are programmed to suffer if we transgress or if we are not moving on further along our pathway.

Fear

How does fear arise? It is based on the anxiety of not knowing how things will turn out. Where there is knowledge, there is no fear. Fear thrives on ignorance, but it may be translated into a positive reaction. Fear may make me drive cautiously and safely and thus avert an accident.

At all times we are trying to bring forth what we really desire and dispel what we do not desire. Some of our fears are really progressive because they move us toward realizing our deepest desires. Fears can be translated into action; we may fear being punished for often acting incorrectly, so we change our behaviour. We may fear that if we walk on a ledge in high winds we may fall, so we avoid this possibility.

Most fears come about because we are afraid of change. Fear has been defined as false evidence appearing real. It is we ourselves who strengthen false evidences and give them reality. Fears in our imagination are lies and like all lies, if we face them, they disappear.

If we did not have fears, we would be unable to keep within boundaries. If we did not fear transgressing nature, we would constantly abuse it. Without proper fear and respect creation cannot last. Everything in existence emanates from one source, and the ultimate purpose of everything in existence is love. Hence, everything is based on love, even fear. It is our attitude to it that determines the end result.

No fear means no movement, no future; no recognition of harmony or disharmony. Fear is a force driving us to be more efficient, more concerned, more aware. If we are accountable, if we take stock of ourselves, we can make our fear more positive. Then it becomes part of the evolutionary process; it moves us toward dynamic growth.

If we fear that at any minute we will die, then we will be urgently spurred on to know the meaning of life now. We will want to know what causes all our barriers to abandonment and stops us from being in true submission to the will of Reality.

Doubt

Doubt is uncertainty and ignorance. We want to escape it because we want to know. Certainty is knowledge. Because we are programmed not to want ignorance, the purpose of doubt is to get us out of trouble and lead us to certainty. If we start with doubt, we will ultimately end up being certain.

Disappointment

*Dis*appointment is a lack of 'appointment'. If you are eager to meet someone but you arrive late and miss the person, you are disappointed. The appointment has not been kept; you have been barred from something in which you had expectations.

Disappointment is a cause of suffering: it is unmet expectations, it is the realization of our fears. What we wanted did not happen in the right manner, at the right time, in the right sequence.

Disappointment is an experience in which harmony and Unity have not been forthcoming. It is a form of shock that awakens us from certain fixed expectation patterns. After our shock of disappointment, if we are living in the 'now', if our certainty and energy are preserved, we will see the perfection of our disappointment. We will see all the elements that resulted in our 'missing the point'. If we let go of our disappointment, it will be transformed into something positive. If we live the moment, every disappointment will become the birth of a new 'appointment'.

The only true appointment is with the One Who is Beyond Time. We are already appointed with Allah. We are the project itself, but we think it is external projects that matter. If we are totally faithful to the present, then disappointment will vanish. So the man of Truth is the man who is constantly appointed; his expectations are met. He is disappointed momentarily but that moment is the birth of the new appointment.

Love and Hate

Love and hate are human qualities that help us to be more discriminating, that give us more proof. We love something because it fulfils us and gives us something we think is necessary to our balance. We think there is something missing in us and that which we love fills that gap and brings us harmony.

Hate is a symptom of disconnection, disharmony, lack of Unity. We hate a person's action because it prevents us from knowing. We hate darkness because we can't see. Love is in Unity and everything hinges on it. That is why we say God's purpose is love. God can never hate; only man can hate, because we have to discriminate what is right, what is wrong, what is acceptable to God and what is not acceptable. We hate what is not acceptable to harmony, ecology, balance and Unity, and we love that which is.

Kinds of Love

There are many different kinds and levels of love just as there are many kinds of hate. The highest level of love is spiritual love – between the Creator and the created. He who encompasses all creation out of love. Allah says, 'I was a hidden treasure and I wanted to be known.' He wanted to show His Essence, His Oneness.

Then there is mother's love, which is biological, hormonal, emotional and human. From one person has come another. A mother's love is basic. The baby crow is the ugliest creature in the world, but when the mother crow was told to go and find the most beautiful bird in the world, she returned from her trip announcing her chick to be the most beautiful.

Man and woman are complementary to each other. Each one possesses something the other lacks. This unifies them into a whole. I may be a very insecure man and find a woman who is always very secure and reassuring. So I love her because she fills a need that I think is important.

It is difficult to define what is meant by loving a person. We love an attribute of a person that fulfils a gap in our attributes. A man may have the courage I lack, so I like him because he's complementary to me. The courageous may like those who are weak because he protects them. That is how negative and positive interact, and that is how harmony occurs in this existence. Once we are aware of this dynamic, we can achieve and understand balance and equilibrium.

Anger

We experience anger when our desires are not fulfilled, when there is an obstruction to our wants. Anger is an energy directed at that which obstructs the fulfilment of our desire.

Frustration is a by-product of anger. We can deal with anger and frustration by either giving up our desire or by finding another means for its fulfilment. We can modify our anger if we are aware that it exists. If we witness what is happening at a given moment, the power and force of our anger will be somewhat modified. Then the anger can be redirected, to try to remedy our situation.

Some feelings of anger are rationally warranted and justifiable. Anger against injustice, against violence, against carelessness, against careless accidents, ill health and abuse, are all positive. This kind of anger brings greater awareness, caution and knowledge.

Then there is selfish anger. The anger of greed, ego gratification and wanting to control the world. This type of anger is senseless, fruitless and feeds on itself.

The roots of anger is found in every heart, but it differs in its extent and its direction. Human anger is a given right; the question is whether it is used properly. When I am really angry, anger manifests through me so I become anger itself. That anger has its own momentum. It lacks checks and balances. When anger is at its highest degree, it becomes all-consuming. I become nothing other than that anger, and this type of situation is basically destructive. It wants to destroy obstacles. If, however, one wants to control anger instead of it controlling one, then one can turn the power of anger into a positive force. It can ultimately manifest as just courage.

Covetousness

Covetousness is the desire to obtain something that is not ours, that we cannot have. If, however, we are able to have it, the desire is not covetousness, but greed. We all have a tendency to want more than we basically need. The motivation behind greed can be positive. There is greed for knowledge, greed for good deeds, greed for activity, greed for success, greed for the best for everybody. But there are also negative aspects to greed.

There is physical greed and mental greed. Physical greed has its own built-in limitations; there is a limit to how much food we can eat, or how many material possessions we can collect. What about other forms of greed? There is no end to satisfying the mind or intellect. Only with the advent of wisdom do we realize that there is a limit to what we can do. So greed ultimately has the potential to balance itself.

Pain

Pain is the outcome of two systems meeting in disharmony. A system was in balance, but I have put pressure on it and now it is in pain. What is the function of that pain? Pain is a signal that something is not in balance. Fear of pain is the fear of not being in harmony.

The pain of disunity occurs when hearts are not in accord. The pain of not being in the right company causes us to be constantly on guard because we do not know where the common denominator is.

The pain of being found out results from our attempts to hide something, to lie. The truth will always come out, and the more we have invested in a lie, the more it will be painful to face the truth.

What a wonderful thing pain is if only we face it, if we allow ourselves to dissolve in it so we become pain itself. If it is psychological pain, then it is dispelled; if it is biological, then we can do something about it unless it is the time appointed for our death.

Pain is the most useful tool of nature. Recognition of its source and how it can be overcome can bring us back into the full orchestration of harmony. Without pain, nothing would happen and there would be no life.

Anything that involves paradox or duality is painful because we want clarity and Unity. We want certainty rather than doubt or confusion. The worst thing we can do for our small children, and we do it out of so-called love, is to give them a choice, because they have not yet developed the ability to choose wisely. Certainty has no pain; yet it is the very pain of uncertainty that spurs us on to know.

We generally think of pain in a physical context, but it is psychological pain that is far worse. There is nothing wrong with pain as such; it is how we interact with it that matters. Without pain we could burn our fingers without knowing it. Pain can act as a warning signal that something is not in line and prompt us to take action.

In phenomena such as fire-walking and certain disciplined martial arts, people do not even seem to experience pain. A single-minded concentration or a meditative state takes a person from the zone of time into non-time. In fact, one doesn't even have to go into an altered state of consciousness – even under normal circumstances when we are rushing about and we suddenly catch our finger in the door, we may not feel pain either. It is only later when our single-mindedness is gone and our mind becomes aware of other parts that we realize our hand is hurt.

In one sense, pain is like disappointment. The moment we get over the shock of it and are able to reflect upon it, it is diffused. Then we see its positive aspect and we can live with it.

Pain, suffering, disappointment are all interconnected. They are all different facets reflecting on the one hand disharmony and disjointedness, and on the other hand Unity and harmony. They are trying to get us to avoid the edge where boundary transgression occurs.

All of the emotions and feelings we have mentioned are divinely inspired. They push us toward ultimate, divine abandonment into the sea of total harmony in time and beyond time. They bring us into true submission to it, but with knowledge and experience.

Arrogance

Arrogance is a protective mechanism that enables us to hide our insecurity or weakness by thinking too highly of ourselves. The more arrogant someone is, the more we can be sure he is insecure. He gives himself away by his rigidity. A plant that does not sway in the wind will break. Arrogance is an imagined protective shield – it protects other falsehoods.

Fulfilment

Fulfilment indicates a desire or wish that is matched by action. We project our desire outward and insert it into our destiny, into the 'decree'. We define decree as the situation that results from all the laws of nature merging together. We all want to achieve and to unify our will with our destiny. If our desires can be fulfilled within the limits of the laws of nature, then achievement occurs. We don't want failure; we don't want misplaced desire; we want to harmonize our inner with the outer. We want to have our intention matched by action; otherwise our desire remains a fantasy.

Acceptance

A state of acceptance implies that there are no incongruities or anomalies between two sysems; they are in harmony. Acceptance, like achievement, is an aspect of unification. Achievement is a more tangible proof of unity, whereas acceptance implies the unification of the outer and the inner. Harmony, intimacy and acceptance are all aspects of unity. We want to belong because we want to be united. We each play a different role, but if we did not have a role, we would feel left out. The truth is that we already belong, we are already an inseparable part of the jigsaw puzzle of life, in which each piece is as important as any other.

Gatheredness and Separation

We are all gatherers. We want to gather knowledge, friends, useful tools, and so forth. The moment we emerge from the womb and are separated from it, we begin attempting to gather what appears to be dispersed. Material success is the result of one's discovery of the need for personal gathering, for the acquisition of goods. For example, one invents something that everyone wants to have, and in return they hand over their money. If one appeals to the gathering instinct, then one gains access to people's accumulated wealth. We perceive separation in duality, in compartmentalization, in discrimination, but it is our nature to attempt to see gatheredness in separation.

Creation hinges on both gatheredness and separation. The more there is separation, the more we want gatheredness. The more we are in outer chaos, the more we want to consolidate. Outwardly everything is in separation. Our inward state can only have meaning if it is in a gathered and unified state. Human nature and experience encompasses both gatheredness and separation. There is only one of us although we may appear in different roles – as a father, a student, a son. What appear to be different roles are gathered at the source, the essence, of our being.

Power

Power is the ability to exercise control. We all want power because it echoes the act of creation; it mimics divinity. Ultimate power rests in the knowledge of the rules that govern power. It is we who are controlled by our need for air, for food, for sleep, and so on. How do we extricate ourselves from the dilemma of being limited by need, but desirous of power? Our desire for power is reflected in our engines, tools and machines and especially in our acquisition of money.

We may suffer from its wrong use, but there is nothing wrong with power *per se*. There is nothing wrong with electricity, but if we put our finger in an empty socket, we will get an electric shock. Power that is not in harmony with nature can be the most destructive. Nature decreed that if we abuse the land and spray it with too much fertilizer or insecticide, it will be barren. The land reacts to what has been done to it. When nature is pushed to its limits, it recycles in order to rectify itself. There is a limit to the application of power; eventually it can bring negative results.

Choosing not to act, choosing not to use power, can come as a reaction to failure. Or it can be a means to avail ourselves of divine power. If we make ourselves completely powerless, our power then comes from pure Reality. Human power is desirable as an insurance to satisfy present or future desires and needs. Higher power is that of true abandonment – it is close to divine power.

5

The Concept of Justice

There is a causal pattern in the universe that governs the laws of balance; the concept of justice is one of these laws. Nature's causal pattern exists independently of human existence. Its laws of balance – in one sense – having nothing to do with us.

We are the cause of both justice and injustice. But how can there be a natural balance in the universe without the inclusion of human beings in that formula? We are not isolated from the rest of the world. In fact, we are the cause of both injustice and the preservation of justice. We have the freedom to disturb the balance. How is it, then, that intrinsically there is justice, yet we can cause injustice?

Justice prevails only as far as infinite time is concerned, but human perception is not infinite. We may see a man beating his son every day and we may know there is ultimately perfection in his affliction, yet there appears to be no justice in his action. What should our response be to this injustice?

We are the instruments of balance. When the seesaw tilts to one side, we have to put our weight on the other side to restore the balance. If we do not do this, we contribute to injustice. Our knowledge of the mechanics of justice does not absolve us from restoring balance when we can. We are not separate in that sense. Justice is meaningless unless our perception of injustice enables us to participate in the restoration of balance. If we are truly living in dynamic submission, then we will do something to uphold justice. If we do not uphold it, we are irresponsible and unjust to ourselves as well as to others.

Absolute and Relative Justice

Natural Reality controls all the factors that exist in creation. Nature is cognizant of everything that goes on. Because of this omniscience, justice prevails. That is why we say that God is absolutely just. Man can only be relatively just because he is bound to miss many factors. Man, therefore, can interfere with God's perfect justice.

So we can disturb the balance of justice in this life by our ignorance unless we depend upon and are totally submissive to God. If we are submissive, we have the possibility of being God's agents. Then we can judge in the name of God.

However, we judge not by assumption, as pharaohs, kings, caliphs, and despots did, appointing themselves as God's agents on earth. Rather, our judgment approximates absolute justice as did the judgment of kings who were prophets or appointed by prophets. They were able, in their state of submission, to consider every factor because they were no longer performing an analytical exercise. Their judgment was not based on circumstantial evidence or investigation. It was divinely based; it sprang from non-time and was applied in time.

The closest we can come to absolute justice, to seeing from God's point of view, with God's eye, is to have no personal eye, no viewpoint of our own. This is the death of the self. Ultimate and real justice is God's justice, and God encompasses all. That is what human justice attempts to be.

How, then, does injustice occur if God is just? Injustice is our human experience. God has not created injustice. Injustice occurs if we do not acknowledge the justness of God. If we observe, without bias, the events that have happened to us that seem unjust, we will see in them the perfection of cause and effect. We will see that there was no other possible outcome than what happened. That is the justice of God, and God's justice prevails everywhere.

Human justice varies according to circumstances. A father stops his child from eating too much chocolate. He is just in taking the chocolate away from his child. The relative justice (or injustice from the child's viewpoint) of depriving the child of chocolate occurs because the father is acting in the child's best interests. But the child is too young to see the justice in this act. He sees his father giving the chocolate to an adult and thinks he is very unjust, but that is because he does not have his father's knowledge. If human justice is in keeping with the decree of balance, then it is right. If it is not, then it is unjust and only exists to serve man's temporary purposes.

Usury is an example of injustice. Usury is against true justice because a person who has no money for food, clothing or shelter would not be able to pay high interest rates. He has nothing to begin with and the usurer demands that he return more than he has. This practice is very unjust, yet we support the laws upholding usury, call it justice and protect it by secular laws.

Human justice is also relative to social circumstance. Not too long ago in the US, slavery and child labour were considered just.

So the values that govern man's justice change. But God's justice and the values that govern it never change. Our view of human justice changes because man-made justice does not stand the test of time. If, however, we make human laws in accordance with the divine laws (as revealed to The Prophets), these laws will survive the test of time.

Punishment is part of divine justice. It is a reaction to an action. If adultery, for example, is not punished, there will be chaos in the land. Adultery is irresponsibility on the part of individuals because they do not want to be bound by contract which establishes duties and responsibilities. They want to get something for nothing.

This law is a precaution to maintain the balance within society. If we do not uphold it, society will decay, as it did in Sodom and Gomorrah. Divine justice will ultimately prevail, so if we do not establish our laws according to it, we will be in chaos. Reality has already established its bounds and laws and has given man reason and opportunity to test and choose what has already been established and intended for this intelligent choice.

Aspects of Justice

Equality and Equity
It is impossible to treat two people in exactly the same way, for no two people are exactly alike. Twins are not the same. No two sets of fingerprints are the same. They may have the same general characteristics, but no two things are equal. In nature, there is no absolute equality. But nature gives opportunity to all creation in an equitable way.

Every person is given some opportunity, but how that opportunity is used, and to what extent, varies. For example, two men may come to a town at the same time with the same amount of money, the same education, and the same intellectual ability. After five years one of them ends up a millionaire and the other one a pauper. Equity is that each had equal opportunity in principle, but they ended up being very different.

Guilt
Guilt arises when one feels a certain amount of irresponsibility or one has been unjust or acted unjustifiably.

In the final analysis guilt arises when one is not willing to face the consequences of one's action, or one does not want to pay the price of one's actions.

If one faces the truth of a situation, one's guilt will decrease. Every action has a reaction, and when one tries to be clever, something in his heart will nag at him. This nagging indicates that the essence of justice is in his heart.

Punishment

Punishment is the final prize of discord and disharmony. It is the medal one is awarded – the same way one is awarded a medal of success – for having been such a persistent, consistent blunderer and failure.

Punishment is the ultimate result, or reward, of a discordant situation. It is the final manifestation of erroneous, incongruous, disharmonious action. The sooner punishment takes place, the more quickly one can put it right. Or if one anticipates the punishment, then perhaps one would not commit the act. The punishment for overeating is obesity. It is a bad reaction to a bad action. The way nature punishes us for overeating is by adding pounds and inches. It is also part of the perfect balance, showing us that we are overweight. Punishment occurs at all levels – physical, mental and spiritual.

If I am driving on the motorway and the punishment of the studs in the middle of the road upon the tyres does not give me enough of a signal to move back to the proper lane, the final punishment is destruction – recycling. If I have proved unworthy in the ecology of the motorway, then I'll be overturned and probably die.

Nature's final punishment is that it will recycle and take us out of the earthly domain. Our punishment is according to the degree of our discordance. The more discordant we become, the more severe our punishment becomes.

Punishment comes as a result of not complying with the laws of existence. Often we are ignorant of the laws. We don't know who we are; we forget our true identity. We are consciousness and energy, but we forget, so we are punished. Ultimately punishment is based on love. All of the laws of creation are based on love. We may not like this fact because of our personal investment in a situation. We are attached to reaching our destination quickly. But we are punished because we transgress the law that dictates the appropriate speed for the tyre – so it bursts.

Punishment is directly related to awareness. If I am a terrible cook and I am unaware, then I am only punished after the customers leave the table; they never return to my dining room. But if I am more aware, then before I have finished cooking, I will put things right. If I am even more aware, I will not do anything wrong in the first place. So if I cultivate my awareness of the interrelations and balance of all things, I will not need to be punished. I will be travelling the path of Unity.

The more we go off the path, the more events appear to us as punishment. Outer punishment is nature's way of recycling those elements that are out of line. So individuals, families and nations will be punished in accordance with their deviation from the path. The mainstream of the ecological dynamic thrust will absorb the aberrations and recycle them.

Deceit

Deceit is a pretence; it is making something appear to be what it is not – dressing up the false as true. Hence, deceit has an element of injustice in it. The purpose of deceit is to gain something that is undeserved. It disturbs the balance. The chair was not antique, but I was charging the price of an antique for it. Deceit is thus an investment in something that is not real.

There are many degrees of deceit. Some deceit may be for a good and virtuous purpose. If, however, I deceive myself about my motives, deceit takes on a negative aspect. For eventually the deceiver is the one who deceives himself. He is in a state of hypocrisy. Hypocrisy implies inconsistency; it is based on illusions and hence has no foundation. But if I deceive myself into being a patient fellow and if I succeed in my deceit, then I may become a patient fellow. If the wrong-doer wears the mask of the saint, eventually he may become saintly.

A king in India had a most beautiful daughter. Many of the neighbouring princes sought her hand in marriage, but none of them satisfied the king's requirements. There was a young man in the town who was pining away out of love for the princess. A wise man in the town approached him and said, 'If you obey me completely, I will get you the hand of the princess in marriage, but you will have to pay me a fee for this.'

The young man agreed to the proposition. The 'consultant' then said, 'You must grow a beard and wear ragged clothes.' He took the young man up a mountain to a cave where he left him to sit cross-legged amid all the dust and cobwebs. After three or four days, the fellow looked as though he had been there for years, saintly and wise.

Meanwhile, the consultant started spreading stories in the town about the holy man in the cave. Within two or three months his reputation had spread for miles around. People came from far and wide to see him and bring gifts.

Finally the king heard about the new guru and sent an emissary to his cave to ask him what gift he would like. The young man told the emissary that he had no needs. The king was very impressed by his reply and said, 'He must be a real man of knowledge.' So the king went to visit him and offered him his daughter's hand in marriage telling him, 'Everybody wants to marry her but I want to offer her to you because you are the only man who does not want anything.' However, the young man refused the king's offer. The king departed in disappointment.

The consultant was furious. 'Look, I built up this reputation for you and when our objective was on the point of being fulfilled, you changed your mind.' The young man replied, 'When I was pretending, I was given everything. But what if I were real? I want to be real now, for then I will have more than everything. I now want to really take to the spiritual path.'

Honesty

Honesty is matching one's outer actions with what is in one's heart. If one's intention and action are unified, one is honest. Honesty implies consistency, and consistency is an attribute of contentedness, understanding and justice.

Beginning's End

The Prophet Moses and his spiritual teacher, Khidr, were at sea. Suddenly Khidr made a hole in the bottom of the boat. To Moses, this seemed a very irrational act for he had no knowledge of the despotic use intended for the boat, whereas his teacher had knowledge of future events and had, by this action, caused future justice to be done. Moses was honest in his outrage, and Khidr was honest in his action. Both were honest, but with different degrees of knowledge.

The more knowledge we have about a situation, the more we understand the quality of honesty. Honesty, the opposite of deceit, is based on knowledge, for the more we know, the more honest we can be.

Moses did not know what was behind the scenes and was impatient for that knowledge, but the situation was a test of his patience. Patience brings knowledge to the seeker on the path, but Moses was impatient, so he judged. His investigation was not complete; in fact, it could not be complete because Khidr at the outset had instructed Moses not to question his actions. Thus Moses had to be patient. It is impatience for knowledge that causes wrong judgment and hasty action.

Honesty is the quality in an individual that causes him to want to do what is right. He is united with his knowledge and his abilities. There is no separation between what he knows and what he can do; there are no breaks in his system. Ultimately the more knowledge one has, the more just one is, for justice is applied knowledge. The more it is applied, the more just it is.

6
Time and Non-Time

Time

We experience time because of the properties of change and movement. It is the space between two events, the punctuations of life, that give us a sense of time. Movement is an aspect of time. Movement can occur in time, and it makes time experiential.

Human beings want permanence; that is, they do not want time to pass; they want to fix time, yet they know that it is moving. The past is an experience that has gone and has left an electromagnetic trace in the computer of our memory. And the future is something that is not yet in the memory bank.

What, then, is the present? What is the nature of now? The more we are aware of now, the less we are aware of the past and the future, the less we recall the past and have expectations of the future. If we stop our thoughts with all their outer agitations, then we experience a mini-death. We have consciously blacked out and entered another realm. People are attracted to alcohol because it masks the past. There is also less agitation and concern about the future because of the concentration on the present. People say things that they normally would not say. Alcohol acts as a release, bringing them into a greater state of temporarily produced unity.

Ultimately the end of now is the end of time. It is complete stillness – the pre-creational, unified and gathered state. We are already in the now but we do not know it. The reason we want to prolong life is because we have a desire to stretch time. Time is already timeless, but we do not know it because we have not experienced stopping time. We want to have a long life because we want to be in timelessness. But if we truly want to experience timelessness, we do not have to experience anything other than the now. Anything else we can experience is not now because it has gone. To stop experiencing is physical and material annihilation; annihilation is death.

Now is this instant, and it is beyond time. To really experience now is to experience the death of time. Now is the time; the end of it is in non-time. We want to achieve because achievement is a way of stopping time. At the instant of achievement and success, our minds are at rest; mindlessness is timelessness.

We can only experience events in time because there is within us a substratum of non-time. It is like the film and the screen – the film moves but the screen is fixed. We cannot have one without the other; they are two sides of the same coin. Similarly, creation could not occur unless there was behind its apparent movement a permanent, unmoving Source.

Time and Non-Time/Death

The existence of both time and non-time implies a relationship of opposites. We are conscious of wakefulness, so that means there is sleep. We are conscious of time and experience it through outward change. Non-time must exist, therefore, and that non-time zone must have its base within us. We want to get to know that non-time zone; we want to 'stop time'. That is why we love holidays, 'mindless' activities, and anything else that stops time. They reduce time in a sense; they reduce past and future, so they take us into the now. Our nature is to seek to stop time, so the quality of timelessness ascends within us. The non-time zone exists deep within human consciousness. We came from non-time, and we will return to it. But in our current life-cycle, we are obsessed by it. Fear of death is fear of the unknown quality of that non-time zone. Those who have had death or near-death experiences verify this. The many experiences of this nature that are recorded show an amazing change in attitude towards life, and fear of the next life is transformed into an acceptance of it as a real experience.

Relative Time

Time is relative; we experience time as both long and short, and yet clock time is the same. Clock time may tell us that an hour has passed, but if we are in the company of someone we love, an hour goes very quickly. In the company of a bore, however, an hour seems much longer.

Our existence in this life is entirely controlled by time and its quality of relativity. We have discovered in this existence that we cannot experience one thing without experiencing its opposite. Therefore, as we have noted, if we are conscious of time, there must also be timelessness. It is theoretically possible, then, for a person to experience zero time, to experience death. The seed of that experience is already within us. It is not something that we acquire; it is already inherent in us.

Patience

A doctor tells me that it is going to take me three weeks to recover from my illness. He knows more than I about the illness, so I accept the situation and am patient. This is positive patience. Negative patience is reacting passively to unfavourable situations. 'I am patient with my sickness,' I say, but have I done anything about it? If I have not, I am acting irrationally and am worse than helpless.

Death

Death as far as we know is the end of an experience. It is the edge, the middle point of the coin of creation. We cannot have creation without destruction. There is a limit to creation. Everything we can experience is within limits, thank God. What would it be like if we experienced no limitations?

Why are we frightened of death? We see death as our enemy because we do not know what it is. We do not know what non-time is, and we are afraid to leave the known. But death is the release of pain for the one who is suffering (from the uncertainties of time). Those around him who want him to stay are behaving selfishly. They fear either the loss of the familiarity of his presence or their own death, leaving behind that which they know, that which is in time. Thus those of us who can stop experiencing time are far better equipped to accept death. The awakened or realized souls are not afraid of death because they accept it as the ultimate release. Death for them is the door to eternal pure consciousness.

Freedom

Nowadays, we believe freedom is doing whatever we want, within man-made laws. We think we are free in this existence, but freedom to do whatever we want is only one aspect of freedom. How can we live without constrictions? The more ignorant we are, the more we think we are free. Yet all of us have a desire for freedom. What is it we want to be free of? What we are really seeking is freedom from ignorance; we want to dispel darkness. We want to be free of doubt; we want certainty. There is only one thing that man can be certain of in this life – that he is going to die. So what he really wants is to know the meaning of death, and since the closest experience to death is stopping time, we all want to experience non-time.

The reason why people cannot stop time quickly is that they are looking forward to something. They have shackled themselves to their houses, their jobs, their situations. However, since non-time is already inherent in us, we are already free – but we do not know it. What, then, is freedom? It is knowledge, the realization that we are never separate from the non-time Reality. The knowledge that there is one Reality encompassing everything is the ultimate freedom.

7

Decree and Destiny

What are Decree and Destiny?

We know that everything in creation is governed according to a quantified measure. There are clear laws for gravitational pull; time is measured in hours, days, years. Measure is a form of decree, for a decree determines the limits or bounds of creation.

A decree is the visible outcome and result of mathematically balanced physical laws. For example, we have the simple law of gravity. If an object that is at rest has its centre of gravity nudged beyond the table's edge, we can say it is decreed that the object will fall to the ground according to a certain speed and acceleration. This physical law has been decreed, and it is not subject to change. We may define destiny as a decree personalized. It pertains to what happens to us individually.

We cannot change the world, but we can change our interaction with it. This is why we say the will of Reality prevails. The Qur'an says that the light of Allah will prevail even though everyone will try to put it out. This means that eventually we

Decree and Destiny

will know the ultimate truth that there is only Allah and every manifestation is a creation of Allah.

All laws move according to divine decree. The cosmos unfolds according to the way it has been fashioned. Freedom comes from recognizing that one has no freedom as far as these laws are concerned. In fact, ultimate freedom comes from the recognition that we are completely subjugated. We have no option but to follow these laws; if we try to break them, we will injure ourselves.

The freedom we do have is to cultivate our awareness of the divine laws. The man of Allah is the slave of Allah. He knows that he is in ignorance, that there is knowledge, and that it is his job to acquire it. Self-knowledge is already within him; he has only to discover access to it. We are all given limited freedom within a decree for each one of us to awaken to the fact that a decree is absolute. It is not relative. We are all programmed to arrive at the conclusion, through our experiencing intellect, and reason, that we are not free.

Destiny is a personal experience of the decrees. It is our individual interaction with the laws. For example, suppose we drive along a road in our car. The road, the road signs and the car are part of a decree, but how we drive will determine our destiny. If we choose to drive properly, we are safe. If not, we may have an accident. What we experience is subjective. Destiny is changeable, but the laws of creation do not change.

The Qur'an says, 'You will never find a change in the Way of Allah.' Imagine what would happen if every half hour the law of gravity changed. Everything would be in utter chaos.

We have defined destiny as the way we interact within the freedom that is given to us, within fixed laws, within a decree. A decree affects everyone without distinction. It was decreed that if the people around the Prophet Muhammad let him down, both they and he would suffer for it. By their transgressions they were afflicted, and so was he. The Prophet Muhammad was wounded in the battle of Uhud, and he recognized that his wound was a result of the transgressions of his friends around him. He had warned them but they did not understand nor did they obey.

So no human being is separate from or immune to a decree. Decrees are dynamic and reveal Divine Unity. We are all interconnected, and whether we perceive it or not, there is one fibre that holds all creation. A decree exists whether we like it or not, whether we accept it or not.

Our individual strengths, our limitations, our knowledge, our understanding, all interact and help create what we call our destiny. It is decreed that we start from the same root and that we end up at the same root – we are of God and from God. It is this connection that we strive to know. It is part of human desire to want to know more in order to learn what to avoid, to sharpen our intellect. The only way we can escape darkness and ignorance is to move toward light and knowledge. We are all driven, pushed, encouraged, in this direction. It is a natural phenomenon to move from basic knowledge, physical and material, to subtle knowledge. We want to know how physical things interact with each other, and want to develop a fine intellectual appreciation, but the ultimate knowledge is the knowledge of Divine Unity.

Ultimately the safe conduct for our passage through life is issued when we realize that our destiny is rooted in a decree. When this occurs, we are in balance. Then we can begin to travel the path of Unity, the path of self-knowledge.

When we have unified with the overall fibre that contains every other fibre within it, with total knowledge, then we can say that our destiny is unfolding in perfect harmony. We are then in a true state of contentment. We may find everybody around us disagreeable, but we can even see the perfection of that – by seeing the cause and effect.

For every action, there is a reaction. The laws of the universe are orderly, not chaotic. Allah reveals in the Qur'an, 'I do not take away a sign from you unless I replace it with something better or the same.' So if our condition is not improving continuously, some aspect of the equation is not in harmony with the laws of the universe. Why do we all want to better our situation if betterment is not written for us and not feasible?

Freedom and Destiny

What is the relationship between freedom and destiny? If, for example, I am parked and a car hits me from behind, what measure of freedom did I have in this situation?

The limitations on our freedom vary according to our situation. For example, we have no freedom from the law of gravity, or from death, but we can overcome gravity by the use of other powerful forces, such as a jet engine or rocket. We have limited, not absolute, freedom in every circumstance. In the case of death, we have no absolute escape from that experience but we can help in postponing it or speeding it up. Ultimately, it is our lack of knowledge that prevents us from taking precautions regarding undesirable ends.

We contain within us a reference point that is outside time. For that reason, we occasionally see events that will occur and we have premonitions, for the non-time zone contains every event in time. That is why the Prophet said, 'God alone has knowledge of the unseen, and God decides what portion of it to give to His people.' If we knew what was to happen to us, our energy would be dissipated; thus, our ignorance is our protection.

Two major schools of thought on the issue of freedom and destiny have often dominated the discussion of this topic. One maintains that everything is decreed and thus we have no freedom. The second school says that man is free, and has total control and responsibility over his destiny. He can do exactly as he wishes. Actually, the truth lies in-between these two views. The truth is that we have limited freedom, freedom within bounds. For example, we are not free of the need for food, but we are free to choose the quality, quantity and manner of consumption.

One day Imam Ali was leaning against a wall and suddenly he jumped away from it. The wall collapsed. The people asked him if he was escaping his destiny. He told them, 'No, I am following the decree.' Imam Ali knew that the wall would collapse; therefore, he moved to save himself. His knowledge of the poor condition of the wall and his ability to move were part of the decree, and his destiny was for him to escape.

Some people are fatalistic. They do not take into consideration measures, decrees and their own will and interaction. They superstitiously blame what happens to them on luck or God's will. God's will for us, however, is to be successful. God's will for us is to be light-hearted in freedom and abandonment. God's will is His mercy. Allah says in the Qur'an, 'I have written upon Myself, mercy.' Those who recognize God's mercy are in submission. If we do not submit our will to God, we cannot know God's mercy. If we abandon our will and submit to His will, then we can know His will and progress joyfully with it.

God's will is spiritual progress. God's will for man is to prepare us for the next life. Many religious people become fatalistic because they do not blame themselves and their ignorance for their inability to act correctly. So they find an excuse, by saying, 'This is what God wills.'

There is collective destiny, as well as individual destiny. For example, if we live with people who drink and abuse drugs, even if we do not, their behaviour will afflict us. We will be subjected to the outcome of their irresponsibility and our destiny is likely to be bound up with theirs. If we are sensitive, we either have to move or suffer. We pay a price for our connection with others.

·

Economic Freedom

We want to be free of material poverty, including hunger, thirst, and lack of clothing and shelter. Money is an insurance, a means to help us free ourselves of our needs and to fulfil desires.

The desire for material or economic 'freedom' is a basic one. Soon after this need is fulfilled and other freedoms are sought; freedom of thought, religion and so on.

Perfection of the Decree

The laws of nature follow a clearly prescribed pattern. If the outcome of a situation appears uncertain, it is only because there are so many factors interacting that the outcome is complex. If one knew all the factors and influences at play, one would see that there was only one possible outcome, with no choice or alternative. Natural laws cannot produce anything unless they are according to the Creator's measure because His measure is perfect and in perfection there is no choice.

The perfect way is one way. If God is perfect, then His will is perfect. God is free of everything, and yet nothing is free of Him in that everything is contained in God and nothing is free of His will. Likewise, the true master is not free of his slave. The true master is free of nothing, for he has no choice in the matter.

Freedom is a road that ends up in complete abandonment, complete submission. We have no choice, but to submit to the way the world is. The more we grow in knowledge, the more we realize that there is only one choice we can make in any situation. We are people of both inner vision and outer action. As people of inner knowledge, we see the reality of a situation; as people existing in the world, we ask, 'How can we help you?'

We are beings of both non-time clarity and in-time action. If these aspects are combined totally, then we are in harmony with in-time and non-time. According to outer action, the laws of nature are for our survival; according to the inner world, they are for our growth in knowledge of the Divine and Absolute.

The Meaning of Prayer

Man's need for supplication springs from the desire to know how to weave the life model he aspires to into the complex net of reality.

The supplicant prays for harmonious interaction with, and a desirable outcome from, the environment, which has within it the many known and unknown forces at play in the cosmos. Prayer is actually a call for a way to steer one's future course into one's unfolding destiny. If the supplicant has any doubts or fears, they will detract from the perfection of his supplication. The degree of success of the supplication will depend on how certain we are that it will be answered.

There is a body of tradition regarding supplication. The Prophet Muhammad said, 'Call upon God and you are certain to be answered.' In a noble tradition, Allah says, 'I am in my slave's opinion of Me. So let him not think of Me but well.' One of our great masters has said, 'Allah will not accept the call of a heart that is forgetful.'

Another states, 'Know Allah while you are in your ease – He will know you when you are in your difficulty. And if you ask – ask God. And if you depend on someone – depend on God. For the pen has written what is to happen until the Day of Judgment, and the whole of creation does its best to benefit you. But what has not been written for you cannot be done.'

The Prophet has said that supplication is part of the decree. True supplication will alter destiny and it is the ultimate weapon of the faithful believer.

Punishment and Failure

We see punishment as the result of a bad action. Let us see, however, how it can be defined if we don't attach a value judgment to it. Then we may say punishment is the negative result of a particular action. We act; a reaction takes place. If that reaction is good, we say we deserve it and it is pleasing. If it is bad, we say we failed, and we try to analyze it and blame it on something.

If we are discerning about cause and effect, we will understand why the result of an action was terrible. We may even be content with this understanding. We will see that the timing was wrong or that we bought the wrong thing. We didn't know that the market was going to be flooded and we would have all these unwanted goods on our hands. We failed and we deserve the result of our failure because we didn't have enough knowledge.

Failure becomes punishment, however, when we look at a situation from the point of view of a seeker. We use the word 'punishment' to highlight the fact that our intentions may not have been right, or that there's a guiding principle trying to teach us not to repeat the same thing. There is no place for superstition when we look at punishment this way. We can immediately translate it into a positive situation. I have failed because of my greed or my desire for prestige. I deserve what happened. The One and Only Harmonizer is at work.

When I bring a moral value to the understanding of punishment, then I can perceive the positive element in it. I neglected my child; now he is on drugs. I have not been responsible enough but through my acceptance and understanding of this punishment, I can make a positive change in the situation. But if I do not take all the factors of the situation into consideration, then I will perceive punishment to be negative. I may become superstitious and blame my situation on bad luck. In that case one has failed to translate disappointment into a new opportunity and renewal.

If we are in real submission, if we are in true Islam, we can see that what is called punishment is of great value. Its purpose is to get me away from the model I am operating on, to deprive me of making mistakes. For a person of submission, punishment is a tremendous opportunity for a fresh start, a precious gift.

The same thing applies to society. If collectively we act arrogantly and foolishly, we will be punished collectively. So if we are good individuals in a society that is acting erroneously, the resulting failure will come upon us too. But if we are people of wisdom (and the prophets were), we will not see this failure as punishment, but as a natural event. We will see that society as having generated a negative outcome, which created the ultimate catastrophe. People and nature interact. We bring about the ultimate doom if we go beyond the bounds of nature.

Punishment carries both outward and inward lessons for the person of submission. We learn where we have transgressed and which elements we did not consider. Any form of greed, arrogance, vanity or anxiety is likely to blind us to the totality of a situation and to take us out of Unity. From the unified point of view, then, punishment is wonderful. It is like stepping stones toward growth, for failures are the points from which we can learn to succeed.

As we know, the root of everything lies in its opposite. The root of the tree of success lies in knowledge of failure. Without failure, without punishment, there is no experience or wisdom. This does not mean, however, that we should seek out situations that bring us punishment. Naturally we do not do that.

To look at an example, we may say that a train gets punished by a jolt when the lines it runs on are not smooth. If that jolt does not improve the railwaymen's awareness of the track's need for maintenance, eventually a time will come when the train will break down. That will be the ultimate punishment and the train will be ready for recycling. It is no longer fit to be on the track – the result of excessive disharmony.

The same thing is true when a culture completely collapses. Initially the punishments for failure to live in unity are small ones. Ultimately, however, there will be a catastrophe if things are not put right. All our failures are lessons to us for progress.

Many people do not have the awareness to see punishment in a positive way. Even so, they may learn to be more cautious and avoid situations that bring punishment. No one likes to be jolted; no one likes to be punished. Punishment is the trophy we receive for failure, but if we are wise, we will use our failure in a positive way. We will not repeat the same mistakes, and we will learn to keep within parameters so as to prevent recurrences of failure.

Sometimes failure can turn people into cowards rather than into wise men and women. But that is not so for men of submission. For them all experience is positive because they are in Islam, in Unity.

If we consider affliction to be good, when we are not in submission, then we fall into perversion. Perversion is separation and duality instead of Unity. Separation prevents us from remembering that we will return to dust, that worms are waiting to devour us. Imam Ali said that when man leaves this world, he will ask his money and goods, 'What have you done for me?' They will say, 'We have clothed you.' And he will ask his kin and friends, 'What have you done for me?' They will say, 'We have laid you in the grave. This is all that we can do.'

If a man remembers the final abode, then his attachment to things takes on a different nature. His respect for them takes on a different form. His relationship with them takes on another spirit. A man of submission accepts the fact that he is going to die, any minute. Our Master Ali Zain Al-Abideen said, 'Allah, make me never forget that the breath that goes in my chest may never come out. Do not make me arrogant enough that if I lift one foot, I assume the other is going to follow.' This is the true man of total awareness and submission. If we talk about Unity then we must love men of truth. If we love God, then we will love the prophets and follow them.

But we do not always do that, for we have subtle hypocrisies. We want to maintain our romantic notions. We want to put Jesus Christ in statue form, or keep him on a gold chain on our chests not in our hearts. We do not live as though any moment Muhammad will walk into our house. We want to pass our responsibilities on to men of school or knowledge – healers, psychiatrists, doctors or others. We forget that each one of us is responsible and that we cannot shirk our duties without paying a price.

According to Islam, each one of us is a leader; we each must know how to lead ourselves. But it is not just the individual who is responsible; it is also the community. People have to grow together. It is not possible to have two members of an orchestra perfecting their craft and the rest lagging behind. If this happens the orchestra will never play together harmoniously.

Likewise, a husband and wife must evolve together spiritually. Otherwise they will not be fit company for each other. The couple must grow together. Similarly, the individual and the community must evolve together.

Occasionally someone will attempt to distinguish between punishment and failure by saying that punishment includes an element that failure does not include. One may feel that in a situation that brought punishment one had an intimation of failure, of choosing the wrong course of action – whereas in mere failure one did one's best, with absolutely no intimation that one might be doing the wrong thing. However, this is a false distinction; to a person of knowledge, every failure is punishment, and there is no way out of this truth.

Every element of disturbance is a punishment, whatever that disturbance may be. Mental disturbance is a punishment, because we are attached to something. We had certain expectations. Anything that takes us away from peace is a punishment. Anything that disturbs our total inner abandonment, freedom and joy is a punishment, for it indicates that we committed an error that has brought on our present state of affairs. Everything that takes us off the path of inner bliss is a punishment.

If we look at the statistics on drunken driving, we will see that many thousands of people are killed on the road each year and many more are injured in car accidents. That is enough to deter anybody. Once we take the risk, we may be asked to pay the price. We cannot be superstitious about our failures, our punishments.

There are situations in life, however, when we will be wronged, when we will be subjected to man's injustice, when we will be treated unfairly, simply because we are in an environment that is not in Unity. In such cases we are afflicted because of the nature of the community, the collective, the environment. Then we must realize that we are not being punished by God but by men who are ungodly. We are merely being afflicted by the state of others and are suffering from the occupational hazard of being human. The prophets suffered in this manner, and on some occasions so will we.

In such situations, however, we may discover the limits of our strength and of our humanity. Then there may come a time when we can no longer stand the environment we are in. Such was the case with Noah. He reached a point where he had to take to the safety of the Ark. A gardener has to have reasonably fertile ground. If he does not, he must move on to another land before he can create his garden.

Our Western society is currently in a situation peculiar to our age. There has never been a culture so rich in facts and information, with so many scientists, educators and artists, with people who have the power to look toward the future and reflect upon it in a detached way. We have all this potential for knowledge, but we debate it and analyze it rather than apply it. If we do not learn to perceive more clearly, if we are not jolted into positive action, then we shall receive one of the most severe punishments the world has ever known collectively.

8

Body, Mind and Intellect

The human system of body, mind and intellect are the mechanisms for receiving, processing and transmitting stimuli. The mind is the seat of our emotions. Animals have minds, but what distinguishes a human being from an animal is intellect. Intellect is the light that illuminates the mind. How do we know that we are emotional? Because we have an intellect. The more the intellect is exercised and developed, the more the emotions are curbed and the mind is put to proper use.

There are two factors that motivate us at all times. We want to open the doors to that which we like and close the doors to that which we do not. We like the familiar because it gives us the illusion of continuity. Every act we perform, then, is based on taking what we want and avoiding what we do not want. The body wants immediate material comforts, the mind likes circumstances conducive to emotional well-being, and the intellect frequently desires stimulation. We are always trying to keep the body, the mind and the intellect in balance with the outside world in unified ecology.

Once our base needs are satisfied, we turn toward the satisfaction of subtle needs until we finally reach the stage of wanting total peace. We want to meditate, to reflect, to sit alone, to listen to the wind and the birds. It is a natural progression, evolutionary and unavoidable. First we struggle for food, clothing and shelter; then comes our spiritual search.

Once man stopped being a hunter-gatherer, the advent of the prophets and their message began to increase. One after another, thousands of prophets sang the same song. They declared that the balance is already here, but it is up to us to stop disturbing it. Knowledge of tranquillity, balance and equilibrium is inherent in us. We know what is right and wrong and we will know the one and only Just Hand provided that we are first of all just to ourselves. But the more we try to match the desires of mind and the intellect with the outer world, the more we will find the futility of it and the more happiness will elude us. A man may retire to his dream retreat in some resort when he is in his seventies, only to find that he has rowdy neighbours.

There are five main systems within the body: the reproductive, digestive, nervous, respiratory and circulatory systems. There are also five senses that connect the body, the microcosm, 'I', with the macrocosm, the outside world. We are always trying to keep the microcosmic and the macrocosmic in balance, in harmony.

We start our journey in this life by developing the body, then the mind, and finally the intellect. Ultimately what we are concerned with is the higher intellect. The child is concerned very much with its physical state. If he is hurt physically, he screams – that's all that matters at that stage of development. But later on our higher consciousness and its demands predominate.

There are different degrees of consciousness. Basic intelligence, physiological intelligence, enables us to look after our bodies. Our mental intelligence tells us how to govern our emotions, and higher intelligence enables us to see clearly in a universal sense and to perceive and seek perfection of everything. We can call it by other names, but this highest intelligence is pure consciousness. Pure intellect is the nearest to the Creator, because there is no ego in it. We see with the eye of Reality. We see cause and effect simultaneously in its perfect mathematical balance. If we live by pure consciousness, then our actions will be in correct balance and efficiency.

The noble tradition says, 'The man of Trust sees by the eyes of Allah, and his hand is the Hand of Allah.' There is no 'self' in this kind of sight; it is pure. It is insight, scientific and without superstition. If the 'self' is dead, Allah is alive, as He always was, is, and will be.

The Five Senses

The five outer senses are the means by which we connect with the physical world. Through these antennae we receive outside signals, process them, and then respond. A taste stimulates my tongue, and then I remember whether I have experienced this flavour before and I respond positively or negatively.

We hold in our memory experiences from the past and we try to compare them to our present experiences. Thus we have a reference point for each of our senses. From the moment we are born, we acquire knowledge of environmental factors. But we have within us an inborn, innate reference point that is universal.

Through our senses we understand guilt, violence, anger, fear and other emotions. However, the form and the expression of an emotion like anger may change from one culture to another. Western anger may be expressed by dropping a bomb. Indian anger may be expressed by galloping off on a horse. The Spanish way of greeting is with a strong embrace. The Eskimo's way of greeting is by rubbing noses.

All mankind has potential access to an innate higher consciousness. For example, we all know what love is, although its form may vary. The love of the mother for the child is expressed by feeding it. The love of the horseman for his horse is expressed by training and riding it. Love is the same, but its expression and the way the senses process it is different. Sometimes love is expressed by pushing someone in the back to prevent him from hurting himself. How can we associate a shove in the back with love? Usually we do not, but a situation may arise when the hardest push in the back may show the greatest love because we are saving someone from the edge of a cliff or a fire.

The senses can be deceptive because of our environmental programming – our expectations, patterns and routines. The Qur'an says, 'And we created you as nations and as tribes so that you will know each other.' This means that each of us can ultimately come to the same understanding, and that ultimate knowledge is knowledge of Allah.

The judgment and governance of our senses originates from one reference point, our higher consciousness. If we want to have a very strong foundation for a building, we dig deeper. If we dig really deep, we reach the centre point of the earth, and everyone ultimately reaches the same point. The ultimate foundation of all the senses is our higher consciousness. The senses are only instruments by which we learn higher awareness and move towards the consciousness of Allah.

Life can only exist in diversity. The surface of the earth is very diverse, and the closer we get to the centre, the less visible difference there is. The same model is true as far as our senses are concerned. The more outwardly focused they are, the greater is the diversity of experience. The more inwardly they look, the more constant and stable is the experience.

Sensuality and the Senses

We have noted that at all levels – inner and outer, body, mind and intellect – we want harmony. At the body level, sensual gratification brings about physical harmony. For example, if I am in good health, if I eat well, if I am massaged and pampered, then the resulting harmony brings about bodily tranquillity. Tranquillity is an aspect of contentment. For example, the climax of sexual intercourse brings contentment, for it releases us from mental agitation and activity. We forget ourselves; we become 'mindless'. Our desire for mindlessness is so great that we even take drugs and alcohol to achieve that state. Alcohol reduces our fears and anxieties of past and future. It makes us forget our troubles and failures. It brings us into the present. Thus, sensuality can be a first step towards higher gratification and balance. If it becomes an end in itself then our end is futile and destructive.

Aspects of the Mind

Mind is experienced because of thought. Its reality is thought-flow. Thoughts are like electromagnetic processes, and our thoughts, intentions and actions are interconnected. We have an intention; it is processed by our mind, and it comes out in action. If it does not, the process has broken down and is incomplete. Human beings are out of balance when their intentions, thoughts and actions are not united.

Thought, like and flow, has characteristics of quality, quantity and direction. We may have seen do-gooders making a mess of things by doing too much. Likewise a wonderful river of the purest water may still bring devastating floods. Good intentions alone are not sufficient; it is necessary to harness them with dams. We often hear people saying, 'I did so much work, but I achieved nothing.' Unless the river is dammed and has a proper direction, it serves no good purpose. All the disciplines and technologies that we may subject ourselves to in this world will be of no use if our intentions are unworthy. But we cannot live on intentions; we live on deeds.

Our quality of thought, then, must be uplifted, our quantity of thought harnessed, and our direction steadfast and purposeful, moving toward higher values. We must constantly try to maintain this balance of quality, quantity and direction of thought. Furthermore we will benefit by the company and counsel of like-minded beings.

There are generally two types of thought: psychological, (emotional), and technological (objective). It is the emotional aspect of thought that causes us the most trouble, not the technological. The technological thought is based on factual experiences and can be useful. For example, I may suggest to a friend that he wear clothing appropriate to a particular climate because I know that climate well. I have a particular wisdom based on experience. So what I suggest is valid. My suggestion is a function of the technological mind and it provides usable information.

Whenever we try to use our mind in an emotional manner, it often causes trouble. 'I hate the way you dress,' is an emotional, subjective and useless statement. The aspect of the mind that is not technological, that is the psychological, is reactionary, and reaction is often regrettable. When we initiate action, rather than reaction, however, we become decision-makers and we behave responsibly, in a unified way.

Emotions may arise within us but, in truth, they are not synonymous with 'us', because some faculty in us is witnessing them. The fact that we can say 'I slept well' is proof that something in us does not sleep. The fact that we can say 'I was angry' indicates that something in us does not get angry. The fact that we can see change enables us to know that there is something in us that is stable and not subject to change – a standard of reference. That is the common denominator in us all.

There is a divine spark that continues to pulsate. If we allow that pulsation to take its course along the right route, we each will evolve into a full being and find that life and death are only separated by a very subtle phase. We will come to know experientially that the next life is another version, a freer, better and happier version of this life and that it will be determined by the overall inner state we are in at our death. Our state is not separate from our actions, and our actions are not separate from our intentions.

There are two types of knowledge – technological knowledge and self-knowledge. Technological knowledge is data, facts and information. Self-knowledge is within us but we have to uncover it. We have to remove our assumptions and anxieties and then surrender. If we uncover and expose ourselves we will find ourselves. Everyone has a demon within him; everyone has his share of negative qualities. If we nourish them, they grow. If we subdue them, while nourishing our positive qualities, our positive qualities will grow. This struggle is necessary before final abandonment and metamorphosis.

We all want to balance and relate our body, mind and intellect with the outer world. We want to match them with the events in the world, so that the outcome is unified and completely balanced. If it works, then we will have achieved 'appointment'. If it does not work, we will have disappointment. By persevering on the path to self-knowledge, we may arrive at a point where we may be ready for true inspiration.

Real and pure inspiration comes to us, however, only when we are empty of all impurities. To inspire means to breathe in. But how can we breathe in if we have not already exhaled? Inspiration comes to us only when we are not full of ourselves, our arrogance, greed, whims, discomforts, and expectations. Inspiration comes to us when our hearts are clear. Then true reflection can occur.

True inspiration usually comes when one is on the point of giving up the struggle. The more our struggle is on the right path, the more our mind is in a better direction. The more open we are, the more we are likely to be inwardly guided. In other words, the struggle is mostly perspiration, and some inspiration. Inspiration comes when we are not worried, anxious or expectant. We are acting and not reacting. We cannot be inspired when we are emotional, for inspiration is dependent upon a state of tranquillity, a tranquillity that is based on dynamic interaction and knowledge.

Heart and Mind

The heart is the seat of our attachments and detachments. If the heart is attached, it only sees the point to which it is attached, but if the heart is free – if it is pure, if it is turning – then it is like a radar beam whose scan is limitless. We can say our heart is free if we have high intelligence, close to pure consciousness. We are programmed to be higher creatures. Our ultimate destiny is timeless freedom and bliss.

That is the meaning of the truth, 'Man is the highest of creation.' We are programmed to seek happiness, and happiness is a state of harmony in which there are no desires, attachments or feelings of failure. It is perfect, simple, joyful, spontaneous being – in the everlasting presence.

Our heart enables us to know what disturbs and what quietens our minds. Our intellects harness our emotions and ultimately guide our actions so that we are least injured outwardly and inwardly. At the end, our heart becomes so firmly connected with the 'beam' of reality that will enable us to act totally in accordance with the decree and in utter surrender and harmony. At the end, it will truly come to know.

9

Noble Qualities

The Prophet said, 'I have only come to complete the noble qualities of character.'

Desire

Desire is a response to discontent; thus discontent is a power that motivates us. It presses us on; it pushes us towards action. The prophetic teaching says, 'O man, you are struggling all the time, struggling in order to know your Lord.' So the root of desire, the root of discontent, is a power that is genetically in us. Desire is like a fire; the more we fan it, the more it will burn, and the bigger it will become. It exists in order to take us to the ultimate goal of self-knowledge or to the brink of disappointment and the edge of despair.

Ultimately we will discover that we can never be fully content outwardly, and that desire is a trick of nature to teach us that we can never be fulfilled by satisfying a desire. We will come to the conclusion, then, that there is no end to worldly, material or physical desires. A man of wisdom will then reach a point of giving up such desires. He will only take what is necessary for his practical existence and will not hold on to any fantasy that can divert him from the inner quest.

Humanly speaking, there is no such thing as a state of stabilized contentment. We will be content for a moment when our desire is satisfied, but immediately a new desire is born. There is no final point at which I say, 'I am now content because I have satisfied all my desires.'

We do learn, however, that desires are of different types and natures. We have very basic desires to satisfy our physical needs – hunger, clothing, shelter. These basic desires are healthy and can be satisfied physically and materially. We reach a point where we say, 'I am reasonably satisfied, I am reasonably full, I don't want to eat any more. I have a satisfactory home; it's all right.' Basic desires preserve us; they exist to put us in a state of physical balance. They are not harmful so long as we do not carry them to an extreme. They teach us cause and effect and help us to develop our lower intelligence.

We also learn that we can elevate the intention behind our desires. We can change them from being selfish to being selfless, from being for our own gratification to being for the sake of our family, our children, our neighbour – for mankind. For example, when we are hungry, we can eat in order to sustain ourselves, to be useful friends, members of our family, members of our community, or servants of Reality. Perceived in this way we have lifted a very basic desire to a higher end. Desires never end unless we are desiring that which has no end and no beginning. If we desire the discovery of the Creator then we are acting according to our true nature. The reason desire is endless is because we are constantly being led to give up desiring that which is not worthy of desire.

When we say desire will never be satisfied and that it has no limits, we are describing a natural phenomenon. Of course, it will never be limited because we are created in order to desire the Creator of desire – and He is beyond all limits. We are made to desire knowledge of the Creator of knowledge. Desires have no end because we desire Him who has no end. Allah has no beginning and no end.

Attachment

Basically we resent attachment and yet we manage to become attached. How do we resolve this situation? We get attached to our habits, homes, cars, friends. We are attached, but at the same time we resent it. We discover that man has the habit of biting the hand that feeds him. We want to destroy any attachment. We walk away from it, or find fault in it and break it off, especially when it makes us feel dependent – such as attachment to an employer, a parent, a teacher. We resent it and feel insecure because of that direct dependence. How are we to resolve this situation? On the one hand, we like attachment; on the other hand, we resent it and desire independence.

If we find ourselves resenting attachments, then we must be attached to the wrong thing. That is why we break attachment; it is against nature's law. If we are attached to a human being, we end up attacking that human being. If we are attached to what that human being represents, then it is something else. For example, attachment to a teacher will end up as doubt and disappointment of the teacher and denunciation, unless our attachment is to what the teacher is teaching, to what he represents.

If we are attached to what he represents, then it is no longer personal dependence, and there is no chance of slavery or cult-like attachment to that teacher. We cannot resent him because our dependence does not have anything to do with him personally. If we are attached to what our father represents – protection, compassion, understanding, availability, love, care – rather than to the man himself, then all is well. We can develop and understand those qualities in ourselves. The truth is that we cannot be attached to anyone, except to Truth, except to Reality, for we are already irrevocably attached to God. Allah says in the Qur'an, 'I am closer to you than your jugular vein.' Therefore, we will never be able to remain attached to anyone other than Him.

There is a traditional story in which a man from a wealthy family renounced everything and put on a patched robe and went to the desert. Our great Master Imam Ali came to the family's house and said to them, 'You have built a very good home in this life. Are you sure your home in the next life is as good?' And they said to him, 'We don't know, but one of our brothers is following your example. He loves you and follows your example. He is in the desert.'

The family sent for the brother, who returned. The Imam asked him, 'Why are you in such filthy clothes in the desert?' The man answered, 'I don't want any beautiful robes.' The Imam asked him, 'You are afraid of being attached to them since you have come from a wealthy home, why don't you use your wealth for other people's sake? Running away from wealth is fear of being attached to it, fear of loving it.'

What the Imam meant is that there is nothing wrong with a palace or with a house if it is used to help, guide and serve others less endowed, as a means not as an end. If we are not attached to our possessions, they can become sources of satisfaction and contentment to others – so that others may learn the nature of Reality, of the Creator, so they may progress towards higher knowledge and gratifications.

Prophet Abraham's final attachment, which he recognized, was to his son. He overcame it by his willingness to sacrifice him. Once he was willing to make the sacrifice, he was free of that attachment.

Attachment to knowledge itself is for the sincere seeker one of the most difficult and subtle attachments. It is said that Imam Ghazali, when he was going out to seek knowledge away from the city, took a donkey loaded with books. A robber came at night and he told him, 'You can take anything except my books.' Later on when Imam Ghazali awakened to Reality he had a vision that the robber had been a manifestation of Reality. It was the Master Khidr that came in the form of a thief. 'I want to free you to take away what you are attached to, so that you are free. Knowledge is not in the books, it is in the one and only book that is in your heart.'

Attachment to physical things is easier to break than attachments to one's knowledge, reputation, name, ideas, thoughts or images. All attachments, however, lead to disaster because we really want to be detached. We want to be free, and independent of all. We are made to be true slaves of Reality only – no-one else; that is why it is said Allah is jealous. If we look anywhere else, He becomes jealous of us. This means we will suffer, for all attachment causes suffering unless it is part of our path, unless it is a provision used on our journey. Then it is acceptable; we need provisions. We are not attached to them if we are proceeding on our journey. Then they are fuel – usable, consumable, helpful in reaching the final source which was the original source, the only all-encompassing source.

Collecting

One of man's characteristics is to collect. We collect for many reasons, basically for physical survival. At the physical and material level, we want sustenance and, therefore, longevity and continuance. One of the attributes of God is the *Baqi*, the Ever-Continuous. We love and seek all God's attributes.

Man is naturally guided to seek his origin and his destiny. All of the motives behind his actions move him closer and closer toward the discovery of his Creator.

At a more sophisticated level of collection, man becomes the collector of antiques, or rare items. What is a rare item? It is that which is least common. There are not many like it. We are all enticed by the prospect of finding and keeping a rare item. Many of our old medias of exchange, be it gold or other precious metals, become of value because of their relative rarity. We are driven to look for the rarer and rarer. Allah is the most rare and dear, and His attribute is the *Aziz*, the Precious, the Mighty. He is so rare and difficult to find because, in fact, He permeates all.

God says, 'I am close and will answer the call of whomever calls. So call Me.' That means, 'Desire knowledge of Me. Desire to know Me.' All other desires are subservient to this desire. We are seeking the rare, the old, the antique, because Allah is the oldest of all. Old beyond age, rare beyond rarity and, therefore, always present – closest, nearest, most obvious. To 'collect' Him is not easy, so we collect junk!

Despair

We despair because we haven't really found what we want but we are not willing to back our search with enough energy. Or perhaps we have partly changed our minds and that lack of resolve undermines our goal. Our own needs or our situation may have changed so we despair. Despair is related to our failure to achieve a desirable end-product. It is signalling to us that things are not in harmony. We must either bring a new element of energy or some other new component in order to enhance the possibility of achievement. Otherwise, we must dismember the whole thing.

Despair gives an opportunity to review, to look at a whole progression of events – how we started, where we are, where we are likely to end. It is taking stock. The emotional aspect of despair is of no use, but the factual part of it is very useful. There is no room for despair in a unific life, because if we review the situation, despair – with its mixture of fear, anxiety, desire, attachment, expectations and ignorance – disappears. In pure awareness, in pure beingness, there can be no despair.

Difficulty

What is difficulty? Difficulty can be physical, mental, spiritual or of other types, and it results from disordered priorities – from being attached.

We start life more attached to physical things; then we move to mental things, and so on. And the ultimate difficulty becomes inner and spiritual. Of course, the emotional is more difficult than the physical because it is higher, more subtle. The more subtle it becomes, the more difficult it becomes to grasp. The Qur'an says, 'Hold on to the rope of Allah.' Where is the rope of Allah? It is subtle, and thus difficult.

So difficulties are signals or warnings of conflicts, disharmonies, disunities – of something incongruous for us to review. I must ask, 'Can I overcome it? Do I need the help of someone from outside?'

The earlier the difficulty shows the better. That is why we say that if things start with difficulty, they are likely to end with ease – because we will have put them right early on. If we see all the difficulties in a car the first day, then it is much easier – we can put everything right first. The more pain we have early on, the more we are likely to work it out sooner. Shakespeare wrote, 'All's well that ends well.' The end must constantly be well. The Prophet said, 'You die as you have lived, and you will re-live as you have died.' The sincere seeker will face difficulties, absorb them and sublimate them. They disappear the same way they appeared.

Patience

Anything that is physical or natural has a cycle that requires patience. We must make our plans and be watchful, alert and patient. For example, we must be patient with our children. Patience, persistence and constancy go together; they are ingredients for success, strength and real power.

We are impatient with ignorance, danger, abuse and injustice, and rightly so. Patience is a great virtue in the right place and it is very difficult to attain. Patience is related to the experience of timelessness. For example, if we have planted a tree and we want its fruit, but it is going to take six years, we must learn patience. After six years we can say, 'Now is the time for the fruit.' But our six years is equivalent to having been in the non-time zone. Ultimate patience requires that stoppage of time.

The most patient being is Allah for He is beyond time. When we are patient, we discover that we are taking on an attribute of ultimate Reality. All the attributes of the Reality of Allah are virtues and acquiring them give us power and strength. Patience spares our running around uselessly trying to speed things up. The food needs three hours in the oven, so we must forget it while it cooks and do something else. As far as the product in the oven is concerned, three hours have been nullified.

Intellect

Reflection nourishes the intellect. Once the intellect is fully nourished and awakened, then all events and feelings will be seen as emanating from the merciful Divine Power behind us, pushing us on. But the intellect must rise. By intellect, we mean higher consciousness.

When we suffer, there is a light that makes us aware of that state – the light of suffering. Then there is another light that shines on that light, and that is called intellect or consciousness. The stronger the light of consciousness, the less significant is the light of suffering, or attachment, becomes. One of them is a brilliant light and the other a tiny flicker. The more the intellect grows, the more our suffering and attachments become insignificant, until we will not even consider them or give them significance in our lives. A time will come when consciousness becomes pure consciousness. None of our suffering or attachments will stay with us then we will move on. They will not saturate us. If our intellect is fully developed, we will be free.

Such intellectual development and mutation does not mean

that we do not recognize the beauties or qualities of things. It does not mean we do not recognize the wonderful quality of a good meal, or the terrible quality of bad food. We recognize and we are aware, but our disappointment or pain becomes insignificant in the light of the great joy of pure awareness and awakening to the truth.

Sincerity and Seriousness

Only through sincerity do we experience unity and submission to the one true Reality. Ultimate sincerity is being loyal to our divine origin. If we are truly sincere, then we are innocent of seriousness. To be serious about our sincerity is fine, but if we are serious about our failures, which are constantly occurring, then we will be shattered.

To take the world seriously is to become heavy-hearted. To be light-hearted is never to be attached or take this world seriously. If something bad happens, we should learn and laugh about it. But sometimes nothing is learned – only that we lack knowledge and awareness. The only lesson human beings can really learn is unity. The rest is secondary, and if we take this world and what happens to us seriously, if we take it to heart, we will be ruined.

The ultimate truth is that all of this world is a shadow of Reality. We need to be sincere to Reality not to its vanishing shadow.

Contentment

Contentment is disrupted by agitation, but is restored once knowledge of the relationship between events has become evident. A short or long period of contentment is an indicator of short or long equilibrium and connectedness. It is a gauge of our unific quality of life.

Faith

Faith without knowledge is blind. Faith alone is not good enough; it is only a beginning. It must develop and be founded upon knowledge. We want to be decisive in this life, and if faith doesn't lead us to clear decisions, it has not achieved its ultimate value. Faith will ultimately take us to the point of deciding that we are here only to know Allah. Any other love or any other orientation is doomed, unless it is serving this ultimate purpose.

If faith does not take us to that certainty, it is futile. If faith does not lead us to clarity and certainty, it is not realized faith. It is theoretical faith. If faith does not take us to a point of surefootedness, of knowing exactly what we want and how to achieve it, then it is of little value. Theoretical faith is fine to start with, but it must be practised to be valid and well founded.

Trust

Trust is a bond that enhances interconnectedness. It enhances unific life, because it is the nature of human beings to trust and to love reliable and trustworthy situations and people. Trust and experience are connected. I trust that after winter comes spring. I have lived long enough to know it. This is trust in discernible, demonstrable reality. I trust that if I drive cautiously through green lights, I will not have an accident. This trust is based on experience.

The most trustworthy aspect of the experience of life is its source, its Creator, because He knew how and why creation came about. If we have trust in Allah, that means we have *Iman* in Allah. *Iman* means trust, knowledgeable trust. The root of the word is *Amn*, meaning peace, tranquillity. There is no agitation in it. We start with *Iman*, which means we trust that we will come to the meaning of trust, because *Iman* can be blind. I have *Iman* that something will happen so that I will be saved. I do not know what or when or how, but in time it will be unveiled to me. *Iman* is faith in something hypothetical. It is fulfilled when it is proven. The two words – *Iman* and trust – are very close in meaning.

One must act honestly from where one finds oneself. If one is under a creaking roof and doubts it will stay up, one should move away. If, on the other hand, one is certain the roof will not collapse, one ignores the warning signs and holds on to trust based on knowledge that nothing will happen. Faith is not blind. It is based on conviction, and conviction is not sustained by ignorance or superstition.

Knowledge enhances trust, which in turn enhances certainty, and so on. Similarly, doubt and uncertainty increase mistrust and, therefore, the collapse of any falsehood. Man builds his future on trust. He is basically a trusting animal. The bond of trust increases one's level of connectedness and Unity. Likewise the level of subtlety of trust is increased as the seeker progresses.

Hope

Basically, human beings hope for whatever brings about harmony, peace and connectedness, inwardly and outwardly. All hope is related to trying to stop something that causes disharmony, or to creating a new element that brings about increased harmony. We hope for commercial success so that we have more money; we think money can buy more possibilities of harmony and satisfaction by fulfilling desires and attachments.

Hope and expectation are strongly connected. Expectation is related to that which can be realized; fulfilment of expectation can be planned for concretely. Hope, however, contains an element that is more subjective and cannot be defined as clearly.

The levels of hope vary with its object. Someone hopes for health and physical well being, another hopes for awakening to Reality – beyond the physical world.

Pure Love

Love of absolute truth cannot contain anything else but love. Love of Allah cannot contain any hate. Allah is all-pervading. That is why, if we love humanly for the sake of Allah, that love is elevated. If we love another person for the sake of Allah, then that love can never have hate in it, unless the original promise changes.

If we love another person in order to guide them by that which we have, then even if they leave or denounce us we will not hate them. Pure love is spiritual love; it is a love for Reality's sake, for truth's sake. It is unquestionable love. It is not transactional or contractual; it is not based on need. There is nothing wrong with contractual love, but it also has hate in its opposite side. Pure love has no other side.

If I love someone because he is doing reliable work for my publications, and if tomorrow he decides to be unreliable, then I may hate his action. But if I love him and the publications for the sake of Allah, then I merely see his unreliability and continue to love him (the divine potential in him). If every motive is turned toward the higher, then we will not suffer from disappointment or hate.

Generosity

There are four categories of generosity. One is giving what has been asked for; a second is giving what is needed without being asked; a third is giving more than what is asked; and a fourth is giving what one wants to keep. These are levels of generosity, but why is it that giving brings us gratification? The Prophet Muhammad said, 'Give gifts. It makes the heart soft.' It is agreeable to the giver, and receiver as well, of course.

Generosity implies giving something of value – something we want or could use. If we give something that is of no use to us we are not being generous. It is like going to someone's house and the hostess says, 'Please, you must eat this cake; otherwise we will have to throw it away!' The guest then becomes the garbage collector's helpmate. Generosity brings us contact and connection with others, so it increases Unity; another person's need is the same as our need. It shows similarity. Then we recognize the meaning of both a need and its fulfilment.

Generosity allows us to bring another person tranquillity and also to neutralize our attachments. Detachment, contentment and relationship are all aspects of Unity. They are all attributes of the Creator. Generosity is expansion and it reflects the nature of Allah, the most expansive. Because the Creator fashioned human beings to like His attributes, we are programmed toward being generous. And if we are not generous with ourselves, we cannot be generous with anybody else. The Qur'an says, 'And do not open your hand and give everything you own so that you will regret.'

First we have to take care of our needs. That is clearly not being mean. Giving away what we need is sometimes foolish, for we must all have adequate food, clothes, shelter, and so forth. Higher generosity consists of being generous with our most priceless treasures – our lives, our time. Everything else is replaceable. Our clothes and possessions can be replaced. The ultimate generosity is to make oneself available – to be generous with time, energy and love. But giving away what is needed for our immediate requirements can be irresponsible. All depends on the person's state and degree of awakening and connectedness with the source of generosity.

Spiritual Maturation

Spiritual maturation implies that we are in a total spontaneous state of awareness where there is no past, no future. Spirituality implies eternity, beyond time. We are creatures from non-time acting in time. The more we mature spiritually, the more our intentions and inner awareness guide our actions, the less we are concerned about the physical and material side.

Joy, Pleasure and Happiness

Joy implies tranquillity, contentment and availability. The heart of the joyful person contains ultimate generosity, compassion and affection. The state of joy is based on true abandonment – and true submission with knowledge. Joy is a by-product of higher awareness – the light of the consciousness.

Joy is the final pinnacle, the final outcome of having gone through all the spiritual maturation processes, through knowledge. It is the ultimate fruit of living faith – floating on the boat of certainty, free of all except of the responsibility of knowledge and love.

Unlike joy, pleasure is the momentary contentment that comes if we have balanced the body, mind and intellect with the desirables in the outside environment. Pleasure, in a sense, is 'buyable' – one can 'arrange' it. My body is in disharmony, it is hot, so I buy an air-conditioned house. My mind is agitated, so I construct an environment in which I receive only good news. I arrange to be on holiday, lying around on a beach, or sitting next to a pool in a sunny climate. Pleasure is not based on a real foundation; rather it is based on our trying to be in control of disharmony.

Joy is the spontaneous by-product of abandonment. But paradoxically, the by-product *is* the ultimate product. Pleasure is like a meal we have cooked. We have worked hard for it, but we have just eaten it and it is finished – it is no longer. So we have to wait before we can enjoy another meal.

We can never buy joy. We can only obtain it if we have given up everything else. There is nothing wrong with pleasure, but it is not comparable to joy. Joy is something much higher, beyond pleasure. Pleasure is a doorway, it gives us a taste, but it is second best, a mere reflection of joy. Joy is our real heritage; happiness is our heritage; that is why we are naturally inclined toward them.

Happiness and joy are inherent in us. But joy only comes to us in true submission (in Islam), in dying. The death of our desires and our attachments brings joy.

Surely happiness is a state that we all want to achieve at all times. We can be happy while we are busy, while we are quiet, while we are eating, while we are talking, while we are driving. Happiness is an overriding state that can accompany a variety of situations and states. Can we be happy while we are angry? If we are angry about injustice, and we work to stop it, we can be happy in the process. We are doing something good to bring happiness. But one state we cannot be happy in is agitation for happiness is accompanied by peace, tranquillity and contentment, by quiet and order.

The point is that we are programmed to reach, in this existence, a state of pure awareness, of pure consciousness, so that whatever we are conscious of becomes manageable and meaningful. Normally we react; we have a certain stance; we protect our own reputation, our own expectations. A stimulus comes to us, we almost automatically go through a process of assimilation and then we react.

Sometimes, our reaction is irrational and inconsiderate, and we are not even conscious of it. When we say somebody is rational, we mean they use their intellect, they reflect upon a process rather than merely reacting. When we are conscious of what is happening to us, we are most likely to be able to act, because we neutralize the emotional stimulus, in a sense, by the vision of our intellect, by our awareness. Then we can act appropriately, rather than emotionally or hysterically.

If we insult a person and tarnish his reputation or image he will probably become angry with us. He may start insulting us;

he may scream or kick us and walk away. But if he is reasonable, sensible and not too unstable, after a few days he may come back and say, 'It was not I who did that. It was anger that overcame me.'

What was this anger? It was a reaction. We have threatened someone's investment in his reputation. But if our intellect, our consciousness of what is happening, is present and our awareness is there, we see everything differently. We do not react. Our higher consciousness changes our response to the stimulus. If the light of consciousness is pure, then the response will be more real and true. And if we are in complete submission, if we are in complete abandonment, then our response will be the most real and efficient possible. It stands to reason that the more our consciousness is pure, untarnished by our own ego, self and personality and our own values, the more real our response is likely to be.

Permanence

We want permanence because intrinsically we have originated from a timeless state. The decree is already fixed. The laws will not change; we do not want to change. We love Allah and Allah never changes. We like predictable situations. The world is in chaos because we are in chaos. We see it through our own undeveloped subjective lenses. The world is in a state of perfection from the point of view of the Creator. Nature is perfect; hence creation is perfect. We naturally do not want change because God is the Unchanging and God is in us. One of our noble traditions concerning God says, 'Heaven and earth do not contain Me, but the heart of the believer contains Me.' The permanent is within, but it seems easier to seek Him outwardly, until we get tired of this futile search. The outward is based on duality and dispersion in order to develop our consciousness. The inward contains the reality of oneness and permanence.

10

Rules for the Wayfarer

If the wayfarer examines his actions, he will find that they all fall into two categories. The first category is the avoidance and denunciation of certain things and situations, and the second category is the desire for and exaltation of other things and situations. All human actions are motivated by either attraction or repulsion. But some of the things that are most attractive to us, most human to desire, can cause us the greatest problems if we do not approach them from a sound, rational, point of view.

Traditionally, some religious teachings have warned wayfarers about certain pitfalls, particularly those caused by women and money. Warnings are not against women and money *per se*, but against attachment to women and money.

Symbolically speaking, woman represents that which draws man into the worldly domain. She represents earth, which provides the experiences through which man's discrimination and consciousness develop. When we are attached to earth, we become worldly; it drags us down from the height of our divine origin. The earth itself is pure, but it is man's inordinate desire for it that caused the fall of Adam. The 'Tree of Desire' brings wars, aggression and seemingly irreconcilable differences.

Woman is mother earth, man's stabilizing element. Man's creative tendency causes him to plough the earth, to pursue new things, to hunt, to leave his tent. His destructive urge causes him to destroy falsehood and tyranny. Woman, on the other hand, represents the home base, the tent. Man likes a domestic situation, but he cannot be domesticated. He likes to know that the tent, the woman, the home base is waiting for him, for it is his anchor and stability. He is like a ship that goes out, but with a rope tied to an anchorage. The woman most loved by man is the woman who is waiting for him – faithfully, unquestioningly. The earth waits; it is benign, merciful, beautiful and giving.

Woman, and the attraction of woman, tests whether man will exalt her or the Creator, whom she manifests. Will man fall in love with her, or with Godly qualities in her? Do we fall for the means and forget the source, or do we see the source reflected in the means? Woman is the earth, and without the earth we would not exist; but an inappropriate attachment to and desire for woman can occur if we are not watchful. If we desire a woman for the sake of her companionship, procreation and love, then this desire becomes part of our journey toward the next life. Woman's attractiveness is Allah's plot but it is for our sake. It is to see if we are awake and discriminating. It is to discern whom we are really falling in love with. If we are already in love with our Creator, then that love is reflected in the form of love for other human beings.

Just as woman provides comfort and stability to man, so money provides power, protection and the means to avoid undesirables. Money enables us to put a wall around our castle and to keep out whatever threatens us. We might say that money constitutes the wall around our garden and woman constitutes the fruits and sensuality inside our garden. Without these two, money and woman, there is no movement, no life; but they are only reflections of the deep non-worldly desires within us. If we are beguiled by either, we are destroyed.

Neither women nor money will make us totally happy. If we recognize them as means, rather than as ends in themselves, then we are not trapped and they can help us towards freedom.

Two symbols commonly used in Islamic teachings to picture the lower tendencies of the self are the dog and the pig. Man, being the highest of all creation, contains within him the meaning of all creation. He therefore contains the meaning of the dog (anger and repulsion) and the pig (desire and greed).

The dog has both good and bad qualities. This does not mean that it is not a useful creature, but it has a correct place. Its place is not on a bedroom pillow, but at the guard post at the gate. The pig is an animal that wallows in its own filth and eats anything. If we are aware of the doggyness and piggyness in ourselves, then we are in a better position to avoid them.

We find that in the earliest cultures of the pharaohs, the man of knowledge is depicted riding on a chariot pulled by a dog and a pig. This represents the fact that such a man is in control of his lower tendencies. He becomes aware whenever he feels the bark or anger of the dog rising within him. One who realizes that his anger is a result of fear or anxiety will manage not to express his anger in an emotional or unproductive way. The suppression of his dog-like qualities is a result of his awareness. The moment he feels the dog rising, he frightens it away. When the pig rises with its love of wallowing in filth, its desire to overeat and to overindulge, he frightens it away too.

In Islam the symbols of the stomach and the sexual organs are used to refer to man's lower tendencies. We all know how easy it is for us to be dragged down by our baser appetites. Individuals and societies have become decadent by allowing these natural tendencies to be unduly nourished and encouraged. If we nourish the lower tendencies, we will be unable to nourish the higher tendencies, for we have only a certain amount of energy available. Our energy is like a stream of water. If it is directed towards weeds and swamps, it will feed them and no water is left for the fruit orchard. So we have to channel our energy diligently, making small canals and dams to nourish the orchard. Our belly and sexual organs are the parts of us which we most readily allow to rule us. For this reason, if we want to corrupt someone, it can be achieved most easily through these weak points.

There is nothing wrong with eating or with satisfying our sexuality *per se*, providing we turn its direction and purpose to the higher good. Then we are safe. If we eat to preserve our health, and to nurture our ability to serve, then eating becomes a divine act. But if we eat only to satisfy our impulses, then we are merely animals.

The seeker already contains the truth within him, and God will find a way of taking hold of him, guiding him and putting bonds on him. The tighter those bonds, the stricter they are, the more quickly will emerge that abandonment and submission which bring the spiritual awakening whose seeds were already within the seeker.

The best route for the wayfarer is to follow a guide. It is for this reason that the Prophet is essential. If it were not for the prophets, we would not be able to progress, in the course of our lives, to a situation of abandonment, realization, wakefulness, joyfulness, freedom, and total preparation for whatever is to come in the next experience. Otherwise we might fall into the pitfall of stagnant lethargy. It is like cruising along a river: unless we have a reference point, we do not know whether we are moving or not.

Some of the most insidious pitfalls for the wayfarer are the so-called spiritual paths, which often bring false security. The dangers of spiritual curiosity-seeking and of false gurus cannot be overstated. If spiritual paths are what they say, then surely they go beyond the titles, the outward signs and the ritual into an experience that is transmittable. In other words, they can impart something real to one's life.

If a path is not transformative, then it is not real. It may be a little track, but it is not a path. If we are really hungry for unification, if we are hungry to see the face of Reality everywhere, if we are hungry to see Truth and our interconnection with every situation as it happens, if we want inner knowledge to unfold, then we will learn it everywhere, even from the ants. All of life will be our teacher. The wasp will teach us; the wind will teach us; our own pain will teach us.

There are seven major factors that seem to shape the nature of human beings – factors that influence the make-up of each one of us. The first is the nature of our parents and their attitudes towards each other at the moment of conception. Do the parents love each other and are they both eager for the gift of a child?

The second influence is the genetic qualities and heritage of the parents. The third is the state of the mother during her nine months of pregnancy. Both her inner and outer condition as she carries the child in the womb are vital factors in the formation of identity.

The conditions surrounding our birth constitute the fourth influence. Coming into a world in a cold, clinical, whitewashed place is quite different from being born with loving women handling us and putting us to the breast. Birth is a major event in the shaping of one's nature. The positions of stars and planets at birth also have their influence.

The fifth factor is the nature of the first two years of one's life: the breast milk, the mother, the love, the attention. Our relationship with our environment, with our mother in particular, is crucial. The feeding, the cuddling, the warmth, the noises we hear, the smells we smell, the love we feel, leave an indelible impression on us for the rest of our lives.

The sixth factor is our general environment during the years of growth. The people we are with, the ones whom we see, the friends of our parents – all influence us. Our physical upbringing, our growth, our nourishment, what we learn on our own, our education, our training, our housing, our career, all of these factors are very important to us and help form our failures, our successes, and all of our learning processes.

And the seventh influence in our formation is our will, that indescribable, immeasurable quality that can overcome the most insurmountable obstacles. A person born from unloving parents may be surrounded by disadvantages, but if the spark of spiritual yearning is alive, it can conquer all shortcomings. If we truly want to know, if we are willing to give up everything else, then it doesn't matter how many obstacles from our past confront us. We may have one leg, we may have a multitude of handicaps, but we can still reach bliss, the knowledge of reality.

Nature has its own equitable way, but a lot of the equity is unseen, just as spiritual yearning is unseen. If one's heart really wants knowledge and wisdom, then the root of that knowledge within one will rise and overcome all obstacles.

Life hinges on change, which is based on the imbalance that constantly exists between us and the outside world. Allah says, 'Why do my slaves ask Me for comfort in this life when I did not create it for comfort?' He does not mean that He created life for discomfort; He means that we are going to be in turmoil until we give in and surrender to the One and Only Reality. We must surrender intelligently and not give in to things that are wrong. We have to struggle. Life is based on struggle, but not futile struggle. For example, it is no use trying to pass knowledge on to people who do not want it. We cannot force knowledge upon people; they have to awaken to it themselves. One cannot tell people what they are not likely to understand, because they will turn against one. We have to speak to people according to their degree of understanding.

Our basic personality, our genetic characteristics and whatever we have inherited, will not change. But what can change is our overall attitude. What can change is how we interact with those limitations in the overall movement and destiny of creation. We can be truly and utterly in voluntary submission. Then a negative trait such as anger, once it is directed towards injustice or falsehood, will lead to a virtuous outcome.

Time and change are interconnected. Change gives birth to time. Each day is a new day, and yet each day is similar to all others. Change is innate, for example, in the experience of seasons. Change is a necessary condition of time, and we cannot perceive the passage of time unless there is change.

All change brings about a certain amount of stress because we love reliability, constancy and eternity. Even though we like change toward better things, we still want certainty, yet we are subject to change. Change shows us that we can never control anything. As soon as we have a situation under control, another factor beyond our control emerges. This state drives us to seek the never-changing. It is God's way of making us desperate enough to forsake other things and seek Him.

We have seen that there are certain innate patterns in individual human beings. I may enjoy being with people. Someone else may, in keeping with his more solitary nature, dislike being with people. These inner patterns can, however, be modified with a bit of education, a bit of adaptation, a bit of adjustment and habituation. But can basic change occur?

There are certain basic traits that essentially cannot be changed or modified, for example, an inherited tendency towards violence. Fighting can be destructive, but if it is channelled, if the energy can be harnessed, the outcome can be positive. To fight against that which is not desirable, to fight against vulgarity, illness, crime or other afflictions is positive. We cannot make a lover of peace out of a fighter. We cannot force a fighter to sit under a tree and spend all his time meditating. He is an active man; he is not going to change. We can't make a horse into a frog. We can only make him a positive, useful horse.

We are all essentially spiritual beings, but we are also decaying creatures with lower, animalistic tendencies. As time goes by, however, we will find that it is our higher nature that gives us full nourishment and that our lower nature is never satisfied. There is no end to its appetite. But this doesn't mean we are to deny our lower tendencies. We cannot deny them; rather, we must respond to them, knowing that they can never be fully satisfied. We must learn to control them in a positive way; we must never attack them; we can only sublimate them. We cannot ignore them. How can we forget something that is part of our animalistic nature? Once a desire has arisen, we cannot ignore it.

That is why many Eastern traditions teach that real wisdom is attained through marriage, through child-rearing, through the normal rhythms and cycles of life rather than through false or imposed asceticism.

The man of submission and Unity, is not separate from his circumstances, and the situation he is in has its own checks and balances. It is not artificial; it is real; it is spiritual and it contains everything we need for our next stage of growth. In its own time, the situation will unfold. The best we can do is to be honest, true to ourselves, and sincere.

Wayfarers must take everything in their stride and maintain a sense of humour. All of life is humorous in a certain way; it is a serious joke on us to see whether we realize that the whole thing doesn't matter in the long run. Ultimately we are going to die and this situation, which we consider to be very important, will not last. So we must not take it seriously. We must take honesty and truth seriously. We are no sooner born than we begin to die. How can we take our situation seriously? Rather we should take the meaning of it seriously.

If we are in true submission, we spontaneously know the intention of Reality because we have dissolved our will into Its will. The man of submission is the man of Unity; the path of submission is the path of Unity. This world is in separation, and the believer can fall into separation and become superstitious. Then we can fail without knowing why we failed. There is nothing wrong with failure, however, if we can look back and see the perfection in it; then we can move on. It is said, 'The believer is never bitten from the same hole twice.'

As wayfarers, we will discover that prayer is our source of connection with the one and only Reality; that means we are no longer connected to ourselves and our own desires and expectations. If we leave our expectations and desires, we will be inspired to adopt worthy and real expectations and desires. Whether we like it or not, the way of Reality will prevail.

III
Beginning's End

Introduction

In the previous chapters we have discussed the nature of man and what makes him behave the way he does, what motivates him, what his goals in life are, and what are the keys to attaining his goals. We have seen that he is constantly moving toward death, yet he does not want to die. We have shown the resolution to this paradox, as well as explaining the path of dynamic submission from which all resolutions come. Travelling this path gives access to that inner consciousness that lives inside us, waiting for us to respond to it. Each one of us has within him a deep well from which springs of pure water will gush forth if only we are willing to remove the debris we have thrown into the well. We have filled our inner wells with the stones of our fears, phobias, anxieties and self-images. But if we begin to walk the path of submission in earnest, in time the well will become clean and the spring will issue forth.

We have concentrated, in the preceding chapters, on a discussion of the individual person in relation to all the factors we have mentioned. But we know that everything is interconnected, and now we want to explore how the individual is related to, and becomes part of, the family, the extended family, the neighbourhood, the community, the nation, and the tribe.

How do societies and civilizations evolve, and what causes their rise and fall? What propels a culture to reach its zenith and then to decay and die? What forces and powers govern the fates of nations? And is the individual separate from his culture or his society? Can a person save himself in a corrupt culture without saving everyone else? In this chapter, we shall look at these questions from the point of view of Unity. Just as we have learned how to see the positive and negative in individuals, so we can learn to perceive the virtues and vices of a society and what our proper relationship to that society may be.

In the Western democracies we claim to be truly linked within democracy. We pride ourselves on unity within diversity, on our tolerance of differences. But are we really united? We find people throwing up their hands and blaming 'the government' or 'the system' for their troubles. But who created the system? Just as we reviewed various aspects of the self in Parts I and II, we will look at nations and societies in Part III. We will look at the origin of nations and the positive and negative aspects of nations. We will also examine whether there is a way out of the negative aspects, the dilemmas that nations suffer from. We have discovered that there is a way out for individuals; is that way applicable to societies as well?

We should first explore what constitutes the concept of nation and where this concept originated from.

11

The Wealth of Nations

Economically speaking, there is strength in numbers. Division of labour, for example, has obvious advantages. No individual can manufacture, all by himself, all the products that he normally needs. Imagine if each one of us had to make his own shoes, construct his own home or manufacture his own car! So there is strength and abundance in collectivity. I grow food, she keeps house, he builds bridges, he runs a store. Ten people can have a much higher economic standard of living if each one specializes. This is rational common sense, and man has 'specialized' since the beginning of time.

Another positive aspect involved in people joining together is the feeling of commonality that is developed. The bonds of commonality may initially develop out of economic need, with neighbours helping each other with resources – food, fuel or whatever else is needed. Commonality also develops when people share the same values, the same language, colour or geographic location.

Physical proximity is an important common denominator. The spontaneous, organic grouping of people into clusters is a natural consequence of the way we have been created. Clusters of people care about each other and share their interests, resources and values. Clusters also create strong group checks and balances. Rules of behaviour emerge from groups with strong bonds. Familiarity enhances and increases generosity, compassion and other innate virtues among members. Such is the outcome of a positive community. A community can be considered to be a germ of a culture. A culture develops slowly over a period of time out of the components of communities.

Culture Unifies Its People

Based upon and limited by the knowledge and experience of the people who form it, a culture arises and develops within a specific environment. A culture has many unifying factors: the cohesiveness of the people who form it, their values, the cultures in proximity to them. By its very nature, a culture itself provides an important factor in unifying and harmonizing.

A nomadic culture may have as its economic base two goat skins, a few donkeys and a rug. Or there may be a culture based on skyscrapers and concrete. A nomadic culture merely scratches the earth's surface as far as economic acquisition is concerned; nomads gather just a few possessions in order to subsist. Other cultures, of course, are rooted in – or mired in – the acquisition of multitudes of things.

As a culture develops, progresses and matures, it becomes more static, more homebound, for man seeks stability. Because it is the nature of human beings to seek that which is ever stable (Reality itself) we reflect this search in other activities in our lives. From initially being nomadic, a culture progresses to cultivation, to agriculture, to development of a town, a locale. But at the same time when we seek stability, we also desire diversity of experience. So, as soon as we have put down roots, established a home base, we start to develop mobility.

Outer and Inner Mobility

America and the motor car is a prime example of a culture that has developed material mobility. The major movements of developments of cultures can be traced to aspects inherent in the nature of human beings. Man is mobile because he wants to be everywhere and know everything. Therefore we have invented steam engines, gas engines, jet engines. We have maximum physical mobility, but we do not have the contentment we long for. There is little inner or spiritual mobility.

The West, as a society, is now outwardly mobile but inwardly immobile. The next step is to become mobile inwardly. The negative aspects of Western development are goading us toward that next step of development. Otherwise it will become more decadent and corrupt, and it will end up regressing rather than progressing. But it is precisely because of the negative aspects of the Western culture that many people have begun to travel the path of submission because they have experienced the ultimate material rewards, and these rewards have not satisfied their inner needs.

Western culture now enjoys a higher degree of physical gratification than any other culture in history. We have gathered mountains of surplus food; we have houses and cars, snowmobiles and yachts, jets and computers. Yet we may ask, 'What good has all this accumulation done us?'

Once we have reached the outer limits of materialism and mobility, it is only natural for us to turn toward exploring inner worlds, inner mobility. There is always a pendulum swing when we reach one extreme. Where else can the pendulum go but in the opposite direction? Nature's way is complex: it can recycle corruption and decadence into nourishment and sustenance.

If we believe in the innate goodness of creation, of man and of the earth, then we believe that we will evolve both physically and spiritually. Physically we have already evolved. We are now on two legs and suffer from backaches. But inwardly, we will also evolve; there will ultimately be a spiritual flowering on a large scale for everyone.

The Natural Laws of the Universe

God has given us a certain amount of freedom to see whether we can discover our godliness. If we do not discover it, however, then the natural laws of the universe take over and cultures collapse. All the great cultures of the past brought destruction upon themselves – the Egyptians, the Greeks, the Romans. We, too, will bring destruction upon ourselves if we do not heed the signs of decadence in our culture.

Because we are egotistical, we think we can interfere with nature's ways. We think that we are separate, exempt from nature's laws. We have developed our technology to the highest degree, but we have not established which uses are proper for technology, and which are abuses – the morality of technology, as it were. If the use of our expertise is not based upon Unity, upon the ultimate Reality, then it will be abused. Then people will turn against technology. There is however nothing inherently bad about technology. If the guiding principle behind it is constructive, then it will prove advantageous to humankind, otherwise it will dictate and rule.

More and more we see the development of civilizations without culture. Culture is the fibre that holds civilizations together, and if its fibre is alive, then a civilization will flourish. But if a culture has not developed and matured properly, it has been robbed of its intended foundation, which is unity, then mere outer civility exists, and decadence sets in, without revitalization.

Just as a child must develop physical motor skills before he develops mental and intellectual faculties, so a society or a nation first grows in gross and material ways, becoming more and more sensitive as it develops. But after it has experienced maximum growth, it starts to decline, unless the goal shifts towards the Real.

The Decline

Although any ageing person declines physically, he need not

decline spiritually, mentally or intellectually. Similarly, a culture, a nation, need not decline in its values if it learns to be renewed and changed. Hundreds of years ago, buildings were made of adobe; now they are made of concrete. The physical appearance has developed, but the inner purpose or values have not changed.

Today our culture is in decline inwardly, as regards virtues and real values. But outwardly, physically, technologically, we are very strong. We might compare Western culture to someone who has grown in age but not in intellect or spirit.

If we live, however, according to the path of Unity, we will discover that outer and inner development go hand in hand. The human being is at his zenith in his forties physically, mentally and spiritually. Forty is the age of understanding. The first twenty years are spent in physical development. The next twenty are spent experiencing life situations: choosing a vocation, marrying, establishing a home and family. The third twenty-year cycle is spent in spiritual growth. And as a person grows spiritually, his physical side becomes less important to him. He derives such satisfaction and joy from spiritual insight that, in older cultures, his wisdom is sought after.

The Health of the Nation

Just as we experience personal health, we also experience national health. Personal health is based on the state of one's entire being: physical, spiritual, mental, intellectual and environmental. We cannot separate these factors. That's why we find 'holistic' movements today. They are attempts to interrelate the different aspects of a person. Interrelationship is necessary in the health of a nation as well as in the health of a person. We need to examine the state of our nation. How healthy are we? How healthy is our so-called civilized world?

When we say something is not healthy, we mean that it is out of balance, out of harmony. We have everything we need materially in Western culture, but we are out of balance at a deep level. We have contaminated our environment, polluted our air and water, and shown disrespect to indigenous people. In America, we have killed the American Indians and then romanticized them. We have killed them because they could not be contained or made into 'civilized' consumers, so they were a threat. They stood in the way of progress; they caused inconveniences. So they were destroyed.

The Indians lived completely in the wilderness, completely connected to the natural energies of their environment. Theirs was a culture that could survive only with constant communion and interaction with nature. Theirs was a culture that survived without a roof, and the moment the roof was put on, the Indian culture was finished. It could not move inward inside the house. It could only be inward when it was exposed to outward danger.

The Indian culture was a very specialized culture, very unadaptable. Its spiritual element could not be contained inside a house. Thus, in a sense, it was not the white man who wiped out the Indians; it was their cultural inadaptability. The culture was like an animal that could survive only in a very narrow band of the environment. It was geared to moving up and down the spine of America. The moment the mobility was taken from it, it died. It decayed into drunkenness and dependence.

A Cultureless Society

Most contemporary Western societies have much civil trappings with little deep culture. Now we are civilized but cultureless. We pretend to have culture in the form of theatres, museums and music. We try to make up for the fact that we have no heart, alive and connected to beyond time.

Thus, our civilized society perpetuated the illusion of culture. In the last ten years, we have amassed more information than has been available in the entire history of humankind. Computers bring us more information and less real knowledge, more civility, but less culture. If we veer toward one characteristic, we miss the other. We can't have everything unless we are in balance, and our culture is not. That is why we are experiencing malaise.

Man is living longer today then ever before; but can we conclude we are better or happier? We think that happiness can prevail if we emphasize productivity. We want to be healthy in order to be productive. We prolong life to make more money, not to know the meaning of life. There have been societies in which people have lived for well over a hundred years. Biologically, there is nothing unusual about a lifespan of a hundred years.

So are we healthier now, physically? Our style of life is obviously not compatible with the way we are anatomically constructed. We are not supposed to sit behind a desk or at a machine for eight hours a day. Such an existence breeds boredom, and that is why we have all kinds of gadgets to keep us placated. We invent compensations – countless health spas and exercise gyms, for example. We live a dichotomous life, compartmentalizing work and play. We have a town house and a country house. After eight hours behind a desk, we love to go and chop wood for two hours.

But what is the use of even having good health if we are not happy? Health is a necessary condition, but it is not sufficient. Our outer health may be better, but our inner health is not. We experience more stress and have less time for reflection. We have less time for ourselves, in a sense, than we have ever had, and even that time has to be snatched forcibly from our schizophrenic existence. People 'get away from it all' during holidays. But we have less contemplative time, less time for just being.

Health and Longevity

We like to give health to people. We like to assist with people's health, thinking of the motive or impulse as a humanitarian one. It has its root in our desire to assist and to reflect nature's own way in bringing about a second chance. Long life is only meaningful if it gives people an opportunity to discover the meaning of life, its origin, its purpose and the meaning of death.

An ill person who has not yet awakened to the meaning of existence is helped to regain his balance in the hope that he will use this second chance to reflect and discover the purpose of his existence. Otherwise, regaining health is futile for it can only result in more illness and then death – a vicious cycle. Equally, we may wish someone who is ill long life or good health for our sake, so that we may have the opportunity to discover through the patient the purpose of existence.

The health practitioner's work will only be meaningful if it is based on this premise. It will only be meaningful if it furthers a gross outer healing that promotes inner healing. Witness the high rate of suicides among doctors who do not have a knowledge of inner healing. In England the highest rate of suicides is amongst surgeons.

We bring aid to starving people because we want them to have physical balance and equilibrium. But it will be a losing battle, as it has been throughout history, if it does not lead to its final conclusion, which is to relieve humanity's spiritual starvation. When we begin to provide spiritual nourishment along with physical nourishment, then the loop will close. Eventually, all starvation will be eradicated. We cannot eradicate only physical starvation; it would be incomplete.

The same holds for the situation with refugees. We want to help them escape physical and political oppression, but what about the oppression of the lower self? We harbour them on a physical level, preserving them from outer dictatorship, but we also have to harbour them from their inner dictatorship. We cannot do that to the needy unless we have done it to ourselves. It is really true that charity begins at home.

The Education of the Nation

Let us now take a look at the state of our education. Is it truly connected with our lives? Are we educating and bringing up our children in a truly meaningful way? There appears to be a gap between university education and real life. Thus students experience a big shock when they return to the outside world. They discover that the value system of the university doesn't necessarily hold true in the rest of the world. A student who has studied hard may find a pop musician making far more money than he can make. He may find the job market bleak; there may be no meaningful work for him.

Thus, we discover that our educational system is not in Unity. There is a discrepancy between it and the real world, because we are in disunity. Furthermore, we find that students are being directed only toward materialistic goals and success in this world, with no regard for the education of the individual in the Reality of the other (and next) world. Our education does not connect us to our lives after death. The reason for that, of course, is that there has been so much dogmatic religious education in the past. One of the virtues of science is its ability to point out the nonsensical aspects of a dogmatic religious education. Science and technology have often proved to be a refuge for our thinkers and artists, who have fled intolerable religious dogmatism. Today, however, some scientists are rediscovering religion in the true sense.

The Phenomenon of Unemployment

The phenomenon of unemployment is an invention of the modern age. Years ago, there was no such thing as unemployment. One worked when there was work and one took time off when there was no work. One was essentially employed all the time, enjoying oneself. There was no separation between work and holidays. Sunday was a holiday, spent reflecting upon the 'holiness' of everything. It is impossible for human beings to successfully create false divisions, for everything is interconnected. We cannot separate mental health from physical, psychological or emotional health. They are all interlinked. In fact, the power of the mind is so great that we can cause illness by our lack of incentive to live. If we are disappointed, if we have lost our loved ones, we are more susceptible to illness. We must finally conclude that the health of a nation is only as sound as its unific knowledge and adherence to that path of behaviour. The more that nation or that society lives in Unity, the healthier it will be.

High Technology

In the Western world we live in highly civilized nations that have achieved the highest material standard in living history. We have products that enhance our comfort, we have unprecedented physical mobility, and we have a sophisticated system of communication and transportation.

Our world is far more unified, outwardly, by telecommunication, satellites, jet travel, and international fashions and tastes, propagated by the mass media. The outwardly unifying forces are immense today. We have devised sophisticated means of communication, of transportation, of access to information through our libraries, universities and museums. But along with the creation of constructive means, we have increased destructive means. Opposites always go hand in hand. We now have proton bombs, which we did not have decades ago. We can destroy all life but leave our buildings intact, which shows where our real values lie. We are far more interested in the outer than the inner, so our means are more advanced than our ends.

We hope that by having more means, we will have better ends, but it is not necessarily so. That is why we have a moral problem resulting from high technology. We live with the fear that some despot may get hold of nuclear weapons. If economic means are put in the hands of destructive people, they will be very harmful, just as knowledge about health technology or any other technology is destructive in the hands of people who are not clear about their right use. They may turn technology in any direction. Our means are acceptable only if our ends are clear and right. For that we need virtues which do not change like clothing fashions.

We have seen that happiness is a by-product of a healthy unific life style and that freedom is actually the condition of having no choice. The more a person thinks he is free, the less free he actually is. And the more we think we are free, the more we are ignorant. We think freedom means that every door is open to us. This is a childlike belief, for the man of wisdom knows he can do only one thing, go one way, the optimum way. That means our constriction is greater if we have more knowledge. Then we become most specific and efficient in our undertakings.

Ends and Means

As a society, we have not really looked at our ends. We have little clarity about our goals, so we end up diffused as we are now. We have products looking for markets. We have devised things that are so utterly useless that we have to entice people to buy them. We don't know what the end is, but ultimately we will discover that the end is the beginning. Our end, death (i.e. non-existence), in its subgenetic code, reflects a similar state before our creation. But none of our good means can be put to a proper purpose, to a proper use, unless as a nation, as well as individually, we know where we are going. To have money is to have power, but to be valid, that power must be used for higher purposes. If it is used for purely physical or worldly luxury, it is destructive, and that is the dilemma we find ourselves in now.

If we have no defined end, then our means are unfocused and keep on multiplying pointlessly. The way of Islam is a complete package with defined means and a defined end. But it is organic, not linear. The contemporary, competitive and aggressive Western mind has been trained to be like a conveyor belt. It thinks in terms of cause and effect only, in a straight line. To think linearly is very useful, but if it does not develop the other mind, the higher consciousness, then we live in inner poverty. Our higher consciousness links us to our origins and to life after death, to immortality. If we concentrate only on this mortal life, we will never be content. We will end up continually moving from one house to a bigger one without being fulfilled. There is no end to the desire for acquisition, which seems to be the predominant feature of our society. Material acquisition is good, but only if it is subservient to a higher goal, to the real intention behind the gift of creation.

The Individual and the Nation

What is the proper relationship between the individual and the nation? And which should be given priority? Today we find many individuals blaming society or the government for their ills. Some of us who hold values that are not adhered to by the culture blame the decay of our moral fibre on society or the government. Blaming someone else gives us a false sense of superiority but encourages irresponsibility. The sense of superiority is misleading, for we are all connected. If the mayor of the town is a criminal, everyone is going to suffer. So our blame is artificial; it is a device. We must either help to correct the wrong or leave the society. All of us want to be connected to and united with others in an appropriate way. The individual is like a plant; it cannot survive in a poisonous environment.

Once we recognize this truth, then we feel the need to be with people like ourselves. Birds of a feather flock together. One cannot remain in an environment that is completely incongruous with one's nature, knowledge or beliefs. It won't work. The individual and society are integrally connected. The individual will be benefited greatly by the right government and vice versa.

The Leadership of a Nation

Usually the leadership of a nation is drawn from the largest representative sampling of individuals, but we find that this is not necessarily true in Western democracies. Certain important elements of the community are not represented – those people who want a stricter moral code, for example, or those who believe that there is a divine law that can't be transgressed. There are certain things that must not be allowed, even if everyone votes for them. Supposing the majority voted for a monthly exchange of wives, would we go along with it? In a democracy the majority rules even if it wins on an issue by only a fraction of one per cent. Where should we draw the line? A government is supposed to be a reflection of its people, but what does this mean, practically speaking?

Today people who have a very strong moral or spiritual bent are in a minority. Those who are basically good but don't have any discrimination can easily be bamboozled. Hence the ruling structure is not a true reflection of the people. What constitutes the people? People have within them the capacity for both good and bad, both positive and negative qualities. Statistically speaking, it may be right to say that the government represents the people. Ultimately what is on top must be representative of what is at the bottom of the basket. But what does that mean? In the past, in many cultures, the family tradition was such that the members of the family followed the best among them, the best in intellect, ability, wisdom and survival. Usually the best was the male elder, the father, and if he was not around, it was the brother, uncle or the mother. We are born with the desire for the best. So we want to follow the best. But is our 'representative' government the best? Will it stand the test of time?

The majority decision is not necessarily the best decision. Most of the time we follow our lower emotions because it is easier. It is easier to avoid confrontation, but confrontation may allow blocked energy to flow. Ease at the beginning may lead to more difficulty in the end. Difficulty at the beginning may bring ease in the long run, if one has dealt with a major obstacle at the start. Most governments nowadays appeal to the base and lower tendencies of people to be elected. No government dares to increase the tax on alcohol or tobacco before elections. So the vicious circle grows. Each government ends up representing lower tendencies than before.

Consciousness and Community

In reality both the individual and the government need changing. There is no point in changing the government unless the people change, and no point in changing the people if they have to live in a negative environment created by a bad government. Both are interlinked. But if we discover we cannot survive in a particular society, then, as conscious individuals, we need to find an environment or a community conducive to our values.

Most prophets, along with their followers, left where it was not possible to practise, teach, pray, and pass on the message of Unity. There is only one God all prophets said and taught, and those who would not listen were left behind. The prophets went to people who were more receptive, and a community arose out of these individuals. From the community of lovers of truth came a society. True justice arose organically from the individuals who were transformed. It was really individuals who made the new order, but the ground was more receptive. So the weeds died out and the good plants took over because they believed in Allah and they followed His Prophet.

Thus, those of us who really believe in Unity have no option but to try and enhance our beliefs and practice by being together. If we find ourselves in an environment where we cannot practise our beliefs, then we have to leave and migrate to a situation that is more conducive to our path.

So after we have developed a consciousness of Unity, then we must share our lives with people of Unity. Otherwise our path is mere romanticism. It becomes real only when it is lived daily in community with others of like mind. Then we are able to flourish, our heart becomes more expansive, and we are nurtured by an environment that supports, rather than denies, our values.

An awakened person cannot come into a new environment and be subservient to it unless he pays a heavy price or he is there only to set an example. If one's purpose is sincere and clear, then the outcome will be all right. But in order to do this, we must know whose emissary we are, whom we are representing. We must know our message thoroughly and be able to explain how it is lived. No wise king will send an incompetent ambassador. He will send only an ambassador who knows whom he is representing. So is the case with the King of Kings – the one Lord.

A Holistic Viewpoint

We have discussed the state of the nation, the health of the nation and have seen that much attention has been given to outer health. Now we need to look at the connection between inner health and outer health and the relationship between seemingly separate disciplines.

Not long ago, it was assumed that health matters were purely physical, but now we are learning that unless a person wants to recover, unless he is happy, unless his attitude is right, his body will not regain its health. We are beginning to glimpse Unity, so that a holistic approach, as opposed to specialization, is gaining favour. We are seeing a convergence of other disciplines, of which the mainstream of medical science is beginning to take notice.

We have become aware that modern medicine is often basically palliative, treating the symptom rather than the disease. It is quieter and easier to give penicillin than to deal with the cause of the disease. If the cause is not removed, the symptoms carry on. If our headache results from mental anxiety or insecurity, it will continue until we sort out the insecurity. Fear can be the result of hate, and unless we set our relationships straight, we will end up taking sleeping pills or tranquillizers. A headache, for example, is a warning sign that something is not right. It is not a disease; it is a safety valve. If we destroy the signal by taking aspirin, by cutting the tie to the nervous system which tells us we are not well, then we are failing to take our body's natural alarm system seriously.

In our previous discussion we have seen that as individuals, we have certain attitudes that have both positive and negative sides. Fear is both good and bad. If it motivates us to positive action, then it is good. If it paralyses us and discourages us, then it is bad. In the will of the individual lies the balance between the positive and negative, progressive and regressive, health and illness. This fact is also true of the nation, the tribe, the extended family, the community and the society, for these social groupings are formed by individuals.

A Nation's Greatest Asset – Its People

Let's take a look at what constitutes the positive aspects, the assets, of a nation, and what constitutes the negative aspects, the liabilities. Individually, one's economic assets include material possessions, clothing, houses, money, cars, gadgets, and so forth. A nation's assets are land, natural resources, industry, commercial and technological know-how, and agriculture, but its most important asset is its human resource – the individual, the quality of the individual. It is the individual who makes the choices, implements the decisions, and controls the purposes for which resources are used.

But the worth of the individual depends on the extent of his reasoned submission, because that is what brings him real knowledge. That is, the value of the man is as great as his knowledge of Reality. So ultimately the real wealth and potential of a society is measured by the potential of the individuals within it.

Of course, one of the problems that arises is the lack of harmony between individuals. We are all connected, whether we like it or not, but we do not exist harmoniously together. One citizen may want to build high-rise blocks, to make more money, and another may fight for the preservation of the neighbourhood. So there is constant battle.

The good leader is most helpful in a disharmonious situation. Gandhi, for instance, brought harmony and cooperation to India for a time. He initially united India by pointing out that the British could not solve India's problems, that India was for Indians. But ultimately even Gandhi could not prevent the people from falling into division, for not all people were willing to follow his leadership.

An individual has been given an inner mechanism to guide his actions. If he is truly in submission, will his inner consciousness rule, regulate and guide his life. But society as an entity does not possess a conscience. It requires that there be defined outer rules and regulations, but rules that are based on higher conscience. Rules that do not change with time – only the insignificant (not related to virtues) can change.

Leadership as a Cultural Reflection

A leader may reflect the quality of the individuals in a society, But how accurate the reflection is depends upon a variety of factors. In every society there are people who hold good values and people who embrace bad ones.

In the Western democracies we may ask, 'Does the leadership in this democracy represent the higher, better elements of society?' Democracy basically means the rule of the majority. That usually means that at least 50 per cent of the people have voted for a certain person or policy, but it does not necessarily ensure quality of choice. If one disagrees with the majority vote, one can always leave the community. So democracy is the rule of the majority, and it leaves the minority dissatisfied. But an enlightened majority will not want to alienate the minority, for it is possible that the minority may one day become the majority.

Generally speaking, if one takes a representative sampling of the individuals in a given society, one has a fair reading of the quality of the nation and of the leader to be elected. But this is not always true. Sometimes a society led by a man of Unity is in terrible condition, but the people have recognized their condition and have elected the best person among them. Then again, the situation may be the reverse, with a predominantly harmonious, developed citizenry and a poor leader.

So the leadership of a culture may represent that culture or it may not. It may represent its highest aspect or its lowest aspect. But ultimately leadership ends up shaping a society to some degree. Thus leadership is a critical factor for a society's future. People often believe that if they change a government through a coup or a revolution they will effect positive change. But this tactic rarely works because the leadership that results from a coup most often does not possess the spiritual qualities needed. A society may be united with its leader and still be a bad society. Also a society may not fully agree with a good, enlightened leader, but eventually it may emerge as strong and united.

Majority Rule

What is the source of majority rule and what is the real meaning of it? Majority leadership is elected because it promises to respond to the lowest common denominator. This common denominator usually consists of the satisfaction of base desires. Of course, these desires change from time to time, particularly according to the age of voters. An election in a democracy in which the mean age is twenty-five will bring different results from an election in which the mean age is fifty. Desire for immediate gratification is often more prevalent among the young, whose natures are more impulsive. A younger electorate is more likely to vote according to fleeting desire rather than true need.

The existence of a need implies that a particular attribute or substance is necessary to maintain one's equilibrium and well-being. But the concept of need can easily become distorted. We may think we 'need' alcohol or drugs or a certain level of material possessions. There is a very fine line between what is really needed and what is imagined.

So, those candidates who promise to satisfy majority desires often are elected. These candidates don't necessarily provide the best long-term leadership; they merely have a good feel for what most people desire. Such politicians induce people to vote for them by promising that they will fulfil the desires of the majority.

Thus the leadership often tries to satisfy the lowest common denominator; otherwise the leaders would lose the election. It is common to find within a country a particular aversion to or prejudice against a certain religious or ethnic group. Politicians use such prejudice to their own ends. They win elections by promising to restrict the offending groups.

So it is the leader who convincingly promises to gratify the desires of most people who ends up winning. But as we have already discovered in our discussion of the individual, the more desires we fulfil, the more new ones are created, and a downhill spiral continues. This cycle is as true for the nation as it is for the individual.

Only rarely do we find a leader nowadays who is both popular and wise, who walks the path of submission and is able to raise the spiritual level or quality of his society. Up until early this century it was still possible to find men who were elected because of their high quality and virtues. They were men of justice and knowledge, who served, who did not use their high offices for money or prestige. They honoured the trust and responsibility given to them.

Self-Gratification and the Decline of Society

Following our own desires will only lead us into a cul-de-sac. The laws of the universe are so ordered that mere fulfilment of our own desires becomes self-defeating, for it is unending. When we become conscious of this pattern, when we have cultivated our awareness to the point where we can see this, then we will discard this behaviour pattern.

Just as individuals come singly to this conclusion, in time I trust that nations and societies will collectively come to this conclusion. Currently we are witnessing a rapid decline in the moral fibre of our society because of the effects of lowest-common-denominator-leadership syndrome. But a time will come when the people follow a leader not because he gratifies majority desires, but because he has vision and foresight. Initially people may not like the things he says, but ultimately they will perceive his wisdom. Then revival will occur.

The Idol of Supremacy

Our culture is made up of people who worship supremacy. We applaud supremacy in knowledge, supremacy in power, supremacy in material possessions, supremacy in technology and in the arms race. Why is it that we seek supremacy? We were created to know, adore and worship the Supreme Being; thus our worship of everything that appears supreme is a perverted form of worship. If we place any other values in place of the Creator, then we are bringing disharmony and disorder into our lives. Knowledge, power, money, technology and all material things are only means to an end and if we forget that, we become idol worshippers. Because Western nations have supremacy in so many areas of modern life, we mistakenly think we are supreme. But increasingly time is showing us our mistake.

Religious leaders, educators, authors, artists and many other leading citizens today are trying to make us aware of our decadence and corruption. They are trying to show us that we are living in an age of more quantity and less quality. The present corollary of economic well-being is poverty in the inner life. Competitive and aggressive commercial life which brings material progress often brings spiritual regression. The outer and inner worlds need to be balanced in order for life to be complete.

The Development of Discrimination

It is sometimes said 'The older you get, the worse things become.' At all times, in all ages, in all cultures, including our own, I have discovered that men of knowledge, great teachers and prophets, have observed that the state of things worsens as they get older. Why is this so?

The older a person gets, the wider his perspective becomes. The child's vision is more limited than the adolescent's; the adolescent's is more limited than the young adult's; the young adult's more limited that the middle-aged person's; and so on. The older a person gets, the wider his horizons are. There is a natural progression from playpen to home, to neighbourhood, to town, to state, to country, to world. Depth, wisdom and experience widen one's perspective and parameters. The inner world expands as well. The more discriminating and perceptive a person is, the more he will be aware of disharmony. Anything incongruous, even in a far distant country, will affect such a person. He is more and more aware of humankind because he is increasingly sensitive to interconnections; he is walking the path of submission and Unity.

So this observation does not necessarily mean that things are actually getting worse; it means that one's vision and discriminatory power has increased and one is able to see beyond the limited physical boundaries. The older one gets and the wiser one becomes, the more one perceives things that are out of harmony.

The Politics of Economics – Fabricating Needs

We have talked about the politics of gratification of desire. We have talked about the difference between desire and need. What is the relationship between fulfilling desires and a nation's economic situation? Wealth can be simply defined as the control and ownership of the means to satisfy desires. The more control one has over the manufacture and distribution of a desired item or service, the wealthier one will be.

Some of these desires are real needs: food, shelter, fuel and clothing. But others are contrived or unnecessary needs, fanciful desires, and these desires are fed by advertising, publicity, and peer pressure. Whatever the reason for the desire, the more a supplier can meet its demands, the wealthier he becomes. Whether it is perfume, gourmet food, cars, nylon stockings, or golf courses, if a person imagines he needs it, he creates a market for it; and someone can profit from it.

Imagined or superficial needs create just as real a market as real needs. The worth of many things that our society considers valuable arises from convention or from scarcity. For example, gold is one of the most useless metals. Copper and lead are far more useful. But because gold is a traditional means of exchange, and because people have conventionally regarded it as highly valuable partly because of its scarcity and lustre, it has become the most prized of all metals.

Today Western countries are full of well-groomed golf greens on which we find people rushing around chasing small white balls. A few decades ago, the golf craze did not exist. Now there are millions of people who like nothing better than to spend a few hours on the course. The owner of the golf club or golf course or the manufacturer of golf equipment is wealthy because he is fulfilling a new desire that has been created. A new market makes for new wealth. These desires keep multiplying.

As we have seen, the more we feed desires, the more they thrive. The creation of desire does not of itself make one wealthy; it is the ability to satisfy that desire, to capture the market, that gives one wealth. Thus, we can define contemporary potential wealth as the ability to create a desire and to satisfy it.

Wealth and Consumerism

What we are really saying is that our present economic system is based on the multiplication of desire. Wealth is based on consumption. Without consumers there is no personal accumulation of wealth. So if one wishes to be wealthy, one has to create insatiable consumers. Wealth and consumerism feed each other; the more goods, the more consumers; the more consumers, the more goods. There are natural fluctuations and cycles in consumer-wealth relationships, depending on supply and demand. Greater demand for a commodity brings more production until a new commodity satisfies desires more efficiently or creatively. Also, certain desires reach a plateau and then start to decline.

The so-called advanced societies have more consumers and products than less developed cultures. In particular, televisions and cars are used as measures of how advanced a society is. The more expensive the goods, the more economically advanced a nation is thought to be.

There are, however, other measures of advancement in certain nations. For example, if measured by sheer numbers of cinema-goers, the film industry is patronized to a greater extent in India than in any other country. Does this mean India is more advanced? What it means is that vast numbers of people are trying to escape from the less than satisfying conditions of their daily lives. Often in culturally sophisticated but less economically developed countries, people identify with actors and actresses. Purchasing power is limited, so the cinema provides an outlet, usually lowering the quality of the people by setting bad examples of behaviour.

To examine the gulf between rich and poor, let us look at the basic commodity – water. The American daily consumption of water is 150 gallons per person (1980–85 figures), which is double the amount consumed in Europe. The Pakistani daily consumption per person is only a quarter of that of Europe, whereas in the poorest parts of the world, tens of thousands of people die daily as a result of an inadequate water supply – they suffer both from a lack of water and pollution of available water.

In the United States, the average total daily consumption of commodities like petroleum, gas, coal, wood and cement is approximately four times greater than the amount consumed in other areas of the world with five times the population. In the world's poorest countries, the consumption is a tiny fraction of that of the United States.

As a nation, Americans currently consume daily more than 90 million pounds of meat, 80 million pounds of sugar, and 25 million gallons of coffee and tea. Germany and the United States lead the world in beer consumption. Americans also consume more than 25 million gallons of Coca Cola and other soft drinks daily, and smoke over a billion and a half cigarettes per day.

The advertising industry uses all kinds of gimmicks and psychologically persuasive techniques to market products. Instant coffee was originally marketed by insinuating that a housewife would have more time for her husband if she didn't have to roast, grind and filter coffee. The advertising and media industries also reinforce our desire to be like others and to have what others have, whether it is jeans, a Mercedes, or a Wang word processor.

As the world population increases, the number of consumers multiplies as well. On average, the population of the world is increasing by more than two million people per day. For every person who dies, three people are born. More consumers to be consumed by their excesses.

Side-effects of Consumerism

One of the most insidious side-effects of the production of goods in the West is the tons of pollution generated each day. Literally millions of bottles, jars and cans are discarded daily in the Western world. Thousands of cars are junked, tons of iron and steel are relegated to industrial graveyards, and vast amounts of paper are wasted every day.

Today, many of our measures of success and power are warped. A single painting can fetch a million dollars. A painting itself is not inherently worth that much money; the painting may connote a taste level that is regarded as a sign of success, of 'the good life'. Possessions of this nature have a certain intrinsic value given to them by consumers of a certain breed. It is frequently middlemen, brokers or agents who foster this ethos. If people *en masse* decided that no painting could command a million-dollar price tag, then the market could no longer bear that figure, for people are the market.

Strength through Less Acquisition

Often the more acquisitive a society becomes, the less inner strength it has. Its focus has shifted from inner riches to outer wealth. Such an emphasis finally leads to the weakening of moral fibre. We can see an example of this in the Vietnam War, from which multitudes of American soldiers returned with drug and alcohol addiction problems. The United States soldier in that conflict had to have a lot of material support, whereas his Vietnamese counterpart subsisted on half a bowl of soup or rice per day. It was a confrontation between a moral stand and sheer might. Willingness to sacrifice and be content with little will overcome material brute force.

Multinational Corporations

One of the economic phenomena of our modern day is the

multinational corporation. In many respects, the way these corporations operate is no different from the way smaller corporations or companies operate. But they have the ability to plan more efficiently on a longer-range basis and to conduct market research on a larger scale. They have greater power because of the sheer numbers of people and amount of capital involved.

There is nothing inherently evil in the multinational corporate structure. Only if that structure is used to evil means or ends does it become destructive. Unfortunately an economic entity of such a size often invites manipulation and ruthlessness on the part of its managers. I can create havoc individually, and it can affect many people. What a multinational corporation does is likely to affect millions of people. So we fear multinational corporations just as we fear despots who have access to nuclear weapons, and we fear an organism that, being supranational, is beyond the normal law designed on a national basis.

One of the logical conclusions of the existence of multinational corporations as we know them is slavery to technology. Rather than technology's serving human and humane ends, it becomes a controlling demigod. Because there is strength in numbers and economies of scale, we tend to build large factories and hire large staffs. This trend means we have fewer individual entrepreneurs, small businessmen and women and shopkeepers.

For an example of this, look at the grocery chain business. A few decades ago there was no such thing as a supermarket. Nowadays, because of bulk buying, bulk storage, and bulk distribution, a chain can offer the consumer more goods and services at less cost than small stores. There is nothing inherently wrong with that; it is the abuse of power that comes with quantity and sizeable numbers, and the immoral enticements to buy inferior products, that is evil. Also the customer loses the human contact and personal relationship between buyer and seller, an important emotional need which the supermarket cannot fulfill.

Fast Satisfaction of Desire

A prime example of the trend towards making it ever easier to satisfy desires is seen in the credit card phenomenon. At mid 20th Century this 'plastic money' did not exist. In the hands of irresponsible consumers, credit cards can breed an addiction to goods and cultivate a sense of unreality in money management. Delayed payments, the convenience of carrying a card instead of money, and immediate, large purchasing power all entice the consumer to acquire more and more credit cards. A few years after the credit card craze began, the inflation rate rocketed and so did the incomes of financial executives. But it is really the customer who ends up paying for this economic convenience – in interest rates, in higher prices, and in runaway purchasing.

Because our economic system has such power to create and satisfy desires, it also plays a large part in dictating the social roles of men, women and children, as well as of various age groups in our society. Once the gratification of desire and the stratification of social roles has become entrenched in the lifestyle, it is hard to change. We don't like to change our habits because we like stability. When people get used to a pattern, a name or a product, it is difficult to bring about a change. For example, people still say they are 'hoovering' – meaning vacuum cleaning – because Hoover was the original make of vacuum cleaner. Similarly, 'Kleenex' has become synonymous with paper tissues.

The Misuse of Technology and Knowledge

It is ironical that much of the knowledge that science and technology bring us is not translated into action. We know how bad alcohol is for our liver, we know ice is bad for our stomachs, we know cigarettes are bad for our lungs, yet we persist in using these products and pay a heavy price for the bad habit.

Western nations have the capacity to meet the basic material needs of all their citizens. But so long as there are people in these countries, as well as around the world, who do not have adequate food or shelter, there will never be peace. We are one humanity, but instead of using all our technology and knowledge to care for the ills of people everywhere, we have fallen into the trap of consumerism. We have created desires and aided and abetted greed, and have ignored the hungry and destitute.

Much of what our scientists discover cannot be incorporated into our present systems because they threaten certain important power blocks in that system. So what appears to be a 'self correcting' or 'reviving' situation is not really true. The so called checks and balances are palliative and the reality of 'wealth controls' prevails for the time being.

Certainly one of the side-effects of modern technology is seen in the food industry. Food processing and the use of chemicals and additives are now causing new health problems. Research and information on this topic is still at an early stage, but we do know that many nutrients are destroyed by our convenience-packaging processes. When foods are stored beyond a certain length of time, vitamin C is lost, and when meats remain frozen beyond a specific length of time, the protein content is no longer assimilable. But consumers are frequently blind to the reduced quality of the merchandise they purchase. They see only that they are getting a greater quantity and convenience for less money. We have become conditioned to look for the price tag on the package.

As we have already said, contentment cannot be packaged and purchased. All of our material goods cannot bring us peace or unity. And the more we rely on possessions, the more chained to them we become. We become used to a lifestyle that in turn enslaves us to itself. And the younger generations, who are born into the acquisitive society, start their lives with materialistic assumptions and values.

It is the nature of human beings never to be satisfied by possessions. Ultimate satisfaction comes only with ultimate peace, and ultimate peace comes only with knowledge of the laws of Creation. Only by going back to the source, to the Creator, will we find lasting contentment. Without the spiritual goal our material advancements will only speed up decadence and corruption.

The Man on the Mountain

Once there was a man who did not want anything. He had two robes and he desired nothing else. So he decided to become a hermit in the mountains. But he found that whenever he washed his robes on the river bank, a mouse appeared to chew on the cloth. The man did not want to waste time chasing the mouse because he had come to the mountains in order to meditate. So he went back to the village and said, 'I want a cat because I have a mouse problem.' So someone gave him a nice cat.

Back to the mountain the man went with his cat. The cat caught the mouse and ate it, but because there was no more food for it, the cat became thin and sad. After a few days, the man returned to the village saying, 'I want some milk for my cat. I live on berries, but my cat needs milk.' Now of course the man couldn't keep returning to the village each day for a new supply of milk, so soon he asked for a cow. Of course, the cow had to be milked, and since the man did not want to take time away from his meditation, he decided to bring a poor, hungry man back to the mountain to milk the cow. He told him, 'I have a cow; if you want to come with me, all you need to do is to milk the cow and give some to the cat; the rest of the milk will be yours.'

After a while the poor man grew lonely, so he told the hermit. 'This life is fine for you; you are meditating, but I need a wife; I can't be here alone like this.'

So the hermit said to himself, 'I can't deprive him, the poor fellow.' So a wife arrived, then children came, cousins came, and finally a whole city descended upon the mountain.

So we have no outer escape from the world, for we are the world and if we run away it will chase us. The only possible escape is from its form to its meaning, and man contains both within him.

12

The Debt of Nations

Education and Unity

From the point of view of the person who wants to see how all knowledge is connected – technological knowledge, existential knowledge, self-knowledge – education is connected to knowledge of Unity. Education in the highest sense enables one to see a total pattern; it equips one to make connections and to understand and deal with any situation that may arise. If a person sees the total picture, then he can safely divide it into different segments – technical education, health education, social education, and so forth – without being simplistic or over compartmentalizing it.

Ultimately, the goal of education is to equip a person to interact with the world in such a manner that he contributes to a better understanding, harmony, tranquillity, and the creation of a better environment for human beings. If education becomes heavily weighted toward technical, economic or material ends, neglecting all the other values, then it becomes unbalanced. Then an individual will be able to deal only with that for which he has been trained. He will never be able to develop on the human level or on the moral level or on the level of self-knowledge. Without relating to the other spheres, the specialist will not be able to perform effectively.

Western Values

In our so-called developed world, education is geared to the current values that the Western nations uphold. Those values are connected with wealth, success, power and influence, with the ability to discover potential desires and needs and to fulfil them. It is human nature to want others to embrace the same values we hold. Therefore Western nations engage in training developing nations to accept their attitudes toward education, toward greater quantity and better means without questioning the end. The Western world has convinced itself that it has the best values and the best goods, and even though there is much self-scrutiny on the part of knowledgeable people within Western culture, we have not yet integrated the constructive criticism they have provided.

So we are in the business of exporting our values and our education and our advanced scientific knowledge. But unfortunately the basis of our desire to export is our insecurity. We are anxious that our values be accepted by everyone else. If we were really convinced that our values were really the best, we would not worry about whether others accepted them or not. Time will prove them real.

But our desire to export is also motivated by the natural human inclination to share what is good. If we have a nice house, we like people to come and visit us. If we see a beautiful scene, we like to call others to look at it. Once we become possessive and covetous, then we lose this natural human inclination.

Nevertheless, we do see people who act differently from us as a threat. People who follow the path of submission are often considered a threat because they do not want to be consumerized. They are called fundamentalists and labelled retrogressive and old-fashioned because they do not subscribe to the prevailing normative values. Our materialistic lives are based upon producing more and selling more, rather than on questioning the value of what we are producing.

Instant Gratification

Of course, consumerism, as we have already noted, is based upon instant gratification. There is a false pleasure in having a desire satisfied. Whenever we satisfy a desire, we feel a degree of contentment for a moment. We feel good; we have a glimpse of harmony and Unity; in a minimal sense we are unified. So it becomes a habit for us to have desires aroused and gratified. It is a perverted way of becoming content. We are discontent, then instantly content. So we get the taste of contentment every now and then and become more and more thirsty for it. We become better consumers.

This search for instant gratification by the individual is reflected in the political state and in commercial corporations. They are all after immediate results. Some time ago people were more concerned about long-term results, about their children or their grandchildren or the future of society. But in the last few decades we have found the emphasis switching to immediate gratification. If corporation executives cannot show profits every quarter, they are fired. The same is true of political leaders. They must produce instant gratification.

Aggression

Another liability in modern society is the emphasis on and the validation of aggression. An aggressive person was not looked upon with favour a few decades ago. If a person was called aggressive, he was considered to be destructive and to lack compassion. But not any more. The attributes of aggression and ambition are now considered to be good. We use the word 'aggressive' in all sorts of advertisements. 'We have an aggressive corporation, and aggressive sales force, and ambitious top management.'

People today are very determined to achieve, but they don't question their purpose in achieving. Our values have shifted toward accumulating wealth and power, and we applaud characteristics which for thousands of years were considered anti-social and not conducive to the health of a society.

Artificial Environments

Another tendency of modern society is the segregation of children and older people, away from the 'mainstream'. Many environments are arranged without facilities for children, because we have come to resent children. That resentment is proof of a deep sickness, because we ourselves were once children. How can we resent the truth of what we once were?

Our society is divisive and full of malaise because it is not in Unity, in submission. People misunderstand the meaning of submission. There are large parts of cities in which a child hardly ever sets foot. We have retirement homes where people can pass the last few years of their lives, before they are dumped into the grave with plastic flowers atop of them. Older people in this culture frequently cannot interact with children; all they see is their plastic golf balls – at best!

Similarly, the young are deposited in their depots (day care centres) because it's convenient. Life is much easier when we have special segregated facilities for children. But what price do we pay for this?

Longevity in Western Society

The best resource a culture has is its people, but people with knowledge, wisdom and experience are not being made use of. Recent research has shown the existence of societies in which people live to very advanced ages. Certain parts of Russia, China and North Africa produce people with these very long lifespans. Lack of stress, good air, lots of yogurt and honey seem to contribute to their longevity.

The factor that seems most important in these cultures, however, is the expectations placed upon older people. They are expected to act as leaders; they are needed, and that is what keeps them going. These people feel a purpose to their lives, whereas we artificially separate old people from the rest of the population. The average age in Western countries is higher than elsewhere, but our older people are separated and disconnected.

It all goes back to the situation of the individual. We have seen that the individual is constantly trying to satisfy his desires and expectations, to ward off any element that is not conducive to his well-being. If the individual who is on the path of Unity does not diligently follow the path toward higher values, he will succumb to the lower elements. He must face fear in order to gain courage. Higher values have already been implanted in us by our Creator. It is up to us to follow the path that helps cultivate these values.

Society is in the state it is in because individuals are not encouraging each other toward higher values. In fact, some of these higher values of generosity and goodness are often used in an abusive way. Some of the corporations that give to charity do so for very hypocritical reasons. It can be good for business. Alcohol manufacturers were so successful with teenagers that they considered it prudent to discourage teenagers from going too far!

The Confusion in Values

In Western society, what began as a means to an end has now become the end itself. So we are seeing the end of this beginning. We can no longer survive when we confuse the means and the end. I am not attacking materialism as such; I am attacking the uses to which we put money and the extremes to which we take material acquisition. A certain amount of material is necessary in this life – necessary even to spiritual well-being.

We have seen that the Western world wants to export its values to other countries. To achieve this it initiates a new industry called tourism. We are now bringing disruption and disharmony to formerly quiet and peaceful villages. Tourism usually benefits only those few people at the top of the economic ladder – the industrialists, property developers, banks and the politicians.

There is nothing wrong with travel *per se*; in fact, it can enhance knowledge and relationships. If we travel to a place to broaden our view, and to understand the people and place, then our purpose is worthy. But if we go on a package tour for two weeks and show no sensitivity to the environment or the people, then our travel starts to abuse others, and becomes debilitating to us. It then brings about irresponsibility, not the broadening of horizons.

It used to be that one travelled to a particular location to visit someone in particular and to participate in his or her environment and way of being. One's knowledge and understanding were broadened by such an experience. There is now much less of this type of travel, however. So much is now based on the desire for economic prosperity, a desire that is the foundation of many of the troubles we have in the world, of capitalism and communism alike.

Again we are discovering that our initial assumption and beginning was false and so it is coming to an end. We only hope that that end will not be catastrophic. We are discovering that our attempt to export our values sometimes backfires. Money, food and goodwill are not enough. I have never known an instance of a developed country doing something for an under-developed one that did not result in a bigger cultural disaster. Even with the best of intentions aid programmes don't work unless they are carried out in a unified way, unless they are implemented with and based upon genuine concern for the people who are the recipients. We work in eight-hour units and quantify and computerize everything.

Exporting Western Systems

In the developed world we have brainwashed ourselves into believing we are the best because we are unwilling to face our real failures. We want to have our values emulated by others the world over – even if they are not interested in consumerism and even if they do not want increased individual wealth based on ever growing desires. We continually spawn new 'nervous systems' – the media, communications of all sorts, the banking system, the International Monetary Fund, United Nations agencies, and private voluntary agencies. All of these entities are trying to reinforce the existing system of power in both developed and underdeveloped countries.

What we are doing is encouraging in underdeveloped countries the emergence of leaders who want quick results, so more and more of these nations are falling prey to the international monetary system and the international bankers who encourage the poorer nations to take out more loans on more onerous conditions. Of course these loans are used to buy from the developed world equipment, services, aeroplanes, armaments, pharmaceutical supplies and other goods. This international network is used to foster Western materialism abroad. So the leaders of underdeveloped countries borrow money and find that they cannot repay it. Another agency steps in to make sure the banks do not collapse. An institution originally set up to fund economic development becomes a crutch to ensure that the Western commercial banks do not collapse. Then these underdeveloped nations have to delay payment of loans or restructure them on worse terms than before.

A lot of what we call 'doing good' originates from selfish and material motivation. Many people 'help' others because their conscience bothers them or it is good for their tax position, or business, or public relations. Philanthropy is basically good and it brings about some good, but the real motive is very important. If the motive is good, the philanthropic effort is likely to be of benefit to all; if not, it is likely to be harmful to all in the long run.

Native Cricket – The True Meaning of Colonization

The primary needs of human beings are food and shelter. In cold climates, as in the Northern Hemisphere, the gathering of food and fuel used to be a primary task for survival.

What is the real meaning of colonization, of expanding one's own domain of power? Generally we find that most contemporary 'advanced' cultures have developed in climates and environments that have encouraged acquisition, accumulation and material security. In a warm climate where there was sufficient food and little need for climatic protection, the drive toward accumulation was not nearly so great. Food was available and shelter hardly needed.

So the nations of the North, with cold climates, had to protect themselves. The search for physical necessities dominated people's lives. When Northerners finally discovered the existence of tropical or warm cultures, with their warm, sunny days and generous, trusting people, they found it easy to walk in and take what they wanted. In most instances, they didn't even have to fight for what they wanted to take. The British never really conquered India; they just walked in, and when the Indians realized what was happening, it was too late. They already believed in the superiority of the British and their system of pageantry, railways and cricket.

But along with the outer ravages sustained by those who were colonized, ultimately inner ravages more significant for the future took place. Decadence and corruption ensued and inner values were lost.

The Debt of Nations

After their spiritual strength had been sapped, the conquered started competing with the conqueror and subscribing to his value system. Ultimately the dominated became just as materialistic as the dominator. Once that happened, the conqueror no longer had to fly his flag in the colonized country, for inwardly the country was already bankrupt in terms of values. The people of the conquered land began to adopt the values, the commodities and the lifestyle of the conqueror. The end result is seen in the fact that although cricket is in decline in England, it is flourishing in the West Indies, India, Sri Lanka and Pakistan. These peoples are lagging behind by decades and therefore are truly vanquished, truly conquered, without even any of the side benefits of having a master to toss them occasional crumbs. They themselves are the agents of so-called imperialism or colonialism and yet they are angry about it and blame someone else. They do not realize that the blame stops with them. They need to return to their own values and to benefit from the science and technology of the West and deal with the West, but on their own terms, rather than on the West's terms or with the West's values. These peoples often attack America, but they love the materialism of America. This is indeed neo-colonization.

I call this situation 'Native Cricket'. Those who have been colonized are saturated with Western values and cannot cope with the results because Western values are not indigenous to them. Indians with British values dress up like the British in the oppressive heat of Bombay. This is craziness.

It is the same with football. There is currently less interest in football in Europe than in previous years, yet it is all the rage in the Middle East. All the Arab nations long to have football trainers from the West. Unemployed trainers from the Western countries are paid millions of dollars to train Middle Easterners to emulate this Western form of amusement. This way the despotic so called Islamic governments of the East hope to entertain their people and preoccupy the youth from questioning the corruption and decadence of their dreadful leaders.

The Burden of Materialism

The beginning of what we term the Industrial Revolution had worthy elements in it. The nations involved wanted to produce goods and services that were accessible to all humanity. That was good, but once that became the end or purpose of life, then values became distorted. So the inception of the economic revolution was both positive and negative, and that duality still exists. Today goods and services are widely available, yet we still find a great deal of poverty – even in America and Western Europe. There is an enormous amount of personal wealth in Western societies, but this brings its own burdens and pitfalls. Inherited wealth in particular can be an affliction in that it corrupts more easily. Since the inheritor didn't earn the wealth, he has less respect for the real value of money than the person who earned it.

Positive Submission

We have discussed the path of submission, of Unity. We have talked at length about Islam and gnosis (ultimate self-knowledge). But talking about it and experiencing it are two different things. If a person is talking about sleep, he is not actually engaging in the experience of sleeping. To experience sleep, one must finally go into it. It is the same with the art of submission. A point comes when one has to let oneself go into it and then one is unified with it. Otherwise, one merely talks about the experience rather than participating in it.

The experience of submission brings one to a state that is very different from being at the edge of it – and the edge is very sharp. One can get to the point, or almost the point, of sleep, but then never, in fact, sleep. One can get to the point of almost understanding the true meaning of Divine Unity, but one may not be submerged in it. There is a difference between the butterfly and the caterpillar from which it is metamorphosed.

A Return to Unity

In spite of all the malaise in modern life, in spite of all the predictions of doom, I see for the future nothing less than a return to the way of Unity, and when that happens, it will be tremendous and unprecedented because it will be on a global scale. Then we will really learn the art and meaning of submission and taste the joy and real security that comes with Unity. When people begin to awaken to the path, the whole world will awaken because the world is so closely connected through telecommunications. It is then that we will see the 'net works'.

The original 'net' which underlies all manifestations and causes, belongs to a dimension that can only be experienced subjectively by completely submitting into it. A unification that sees no duality and results in an absolute connection between cause and effect, past time and future time. We all yearn for the knowledge of this fundamental unific source, for this is what works. It is all Allah's work.

13

The Hope of Nations

Types of Truth

There are two types of truth or knowledge. One is capable of proof; it is biological, physiological, existential, mathematical, material. I can prove to someone that if he steps out of a window into the open air, he will fall. It is obvious, natural and provable. It is an undeniable truth that we can demonstrate.

But there is also the truth that contains and unifies with what is called the unseen. The unseen is made up of the forces and powers that we have access to but cannot see or perceive with our outer human senses, like cosmic rays, sound waves, radio transmissions, radar, and so on. These are within the unseen. So there is one type of truth within human grasp and another type that we can tap into by means of other media or connecting systems.

The prophets had access to the second type of truth. What we see of revealed truth is only the tip of the iceberg. Judaism,

Christianity, Islam or whatever other religion we may have been brought up in has reached us in altered form and through different cultural perspectives.

Thus the paradigm we know could be partly true and partly false. Whether the distortions were intended or not, there are certain things that are not coherent in our religious heritage. So what has happened is that these religions, which were complete in their original environment, for those people, at that time, reach us with distortions. Sometimes these distortions are so fundamental that they cannot be corrected.

Thus we find certain elements that do not make sense. We do not connect with them and thus cannot be unified with that religion. As a result we may reject all other revealed truths and deprive ourselves of the proven way of surrender and freedom.

Rebellion against Dogma

What we have found throughout this world over the last few centuries is a continuing rebellion against dogma and despotism, imposition and inquisition. If the heart and the head do not agree, one cannot accept a religion and live by it wholeheartedly. Man wants to believe, but he needs to use both his intellect and his heart to question and accept it.

So science became a refuge for men of imagination and higher intellect, a refuge from the dogmatism and despotism of some of the churches. They simply by-passed its authority. As a result, power has ultimately shifted from the Church to science, technology, economy and finance. The churches' real power has vanished and many are empty. Many churches are either up for sale or have become bingo halls.

Man needs to believe, so he believes in science, which is causality, and cause and effect represent half of the truth. There is nothing wrong with this form of truth, but it is mundane; it is at a lower level. That is why the higher the level of the scientist, the more quickly he reaches a point, described by Einstein as being at the edge of the ocean of knowledge.

There many scientists receive signals, sparks of inspiration from the ocean, and this is basically the state of the man of Unity. He is in that constant state of inspiration. The inspiration comes if you are at that ocean, at that edge, empty-headed, empty-handed, in unison, in resonance.

The Secular and the Sacred

The understanding that in previous ages was accepted as relevant to man, that which clarified the purpose of life and combined the seen and the unseen, the knowledge of life after death with knowledge of this world, is now considered irrelevant. Many scientists today remain in the scientific playpen because of its familiarity and safety. To gaze on the unknown is very disconcerting, for one cannot know what it holds. Nature's trick is every now and then to allow a genuine glimpse into the unseen, and the result is another Nobel prize-winner. But a number of scientific discoveries are merely the result of picking up stray signals – like thieves happening on a cache.

Ever since the birth of modern science, all other areas of man's culture have taken a back seat, and this is particularly true of philosophy and theology. Unfortunately the split between the two types of knowledge – revealed and demonstrable – has led to a split in the lives of people into what we might call the secular and the sacred. All is originally sacred.

Because of the decay of all the major world religions, Islam – the path of submission – has also been considered decadent or backward. One of the beliefs of the Aquarian age is that all religions are one, and so there is no discrimination. But we do not call Islam a religion; we call it the path of unity, of unification – the way of life. But for many people, after the decay of the major religions, God became mundane, a god of causality, a convenient mathematical formula.

The Way and the Reality

The real purpose of religion is to reunify human beings, for we are composed of both a divine nature and a lower nature. The role of religion is to show us that these two natures are actually a part of a spectrum, and it is our decision which side we focus on and develop. Religion is also there to give us the rules and regulations for the highway of mortal life and to make us aware that Reality consists of more than just the houses we live in and the cars we drive. There is a beginning and an end to the highway and they encompass us; we cannot encapsulate them. If we start by wanting Reality, we end up on the Way; and if we start with the Way, we end up with Reality. The two are complementary; they are the two sides of the same coin. We cannot have one without the other. That is why it is said, 'He who takes the Way and does not find Reality is degenerate. And he who takes the Reality and does not accept and fully live in the Way is a heretic. You cannot have one without the other.' One is inner, the other is outer. The two unified are full equilibrium and arrival.

Beginning's End

The moment of truth has arrived, and that is why I have called this book *Beginning's End*. The assumption that there is nothing beyond or behind material reality is false. The moment of truth has arrived because the culture we have built on this assumption is not working. On an individual scale it is not working. On a national scale it is not working. On a global scale it is not working.

The beginning was but a short moment. Therefore we anticipate a hopeful future, but we must have the courage to face the present. The future is in our hands today. The future is a combination of our will, our courage, our understanding, our knowledge, and our past. If these attributes are at a higher level than they were in the past, then the future will be so different from the present that it will scarcely be recognizable. We are a product of the past, travelling ten miles an hour. But if we acquire an outboard motor, the motor of real knowledge and dedication, then our speed will increase to a hundred miles per hour. Our future will differ instantly. All our past mistakes, all our follies will become insignificant compared to our new state and speed of development. If we wake up to the fact that we are created to be joyful and free, that we are inherently divine, then we will wake up from the dream that currently has us in its thrall. The dream is not a bad one, but once the means has become the end, then the beginning has come to its end.

IV
Postscript

Postscript

We have seen that the nature of existence is based on two opposites. We find either increase or decrease, wideness or narrowness, elevation or abasement, strength or weakness, giving or withholding, expansion or contraction, gatheredness or dispersion, happiness or sadness, life or death. Outwardly, life is experienced by duality, moving from one state to its opposite.

The nature of the inward is different from that of the outward. The basic nature of the inward follows along a singular path, rather than leading to a two-pronged outward experience. If we reflect deeply, we will find that the root of these opposites is a single source. If we trace any of these attributes to its root, we will reach a singular point. It is only through the porthole of integrated awareness that the unifying network of the outward can be seen.

We are in this world, yet our higher nature is beyond this world. Our bodies are borrowed and will perish, but our spirit is from a source that is beyond time and, therefore, its nature is permanent.

Thus part of the perfection of the work of Allah is that He

made this existence based on two. He created everything in pairs so that the knowledge of things is by their opposites. By His creating duality, we know that He is One.

This is by the uniqueness of the oneness of Allah, the One, the Unique.

Therefore, the witnessing of that which is unique and has no opposite is so bewildering that it would almost render it unmanifest. Its non-manifestation is because of its excessive clarity, and its deflection is due to its dazzling brilliance.

So we praise the One Who is hidden from His creation because of His intense appearance and Who is hidden from them because of His effulgent light. There is no place or time that He is not. His existence precedes non-existence, and His Foreverness precedes beginning.

A master of our path, Imam Ali Zain Al-Abideen, raised this knowledge of duality to its highest point of worship of the One, as he supplicates:

> O God, make me adore and worship You,
> But corrupt not my worship with self-satisfaction.
> Bring about good toward mankind by my hand,
> But effect it not by my expectation of their gratitude,
> And protect me from vainglory.
>
> O God, bless Muhammad and his descendants,
> Lift me not a single degree before mankind,
> Without diminishing me accordingly in my own eyes,
> And give rise in my soul to an inward shame,
> Of the same measure.
>
> I witness that there is no god, but Allah,
> And that Muhammad is His Prophet.
> No power and no strength except by Allah.
> The Glorious One.'